12-2-75

*HIGHLIGHTS IN*
*Astronomy*

The 4-meter telescope of the Kitt Peak National Observatory.

# HIGHLIGHTS IN
# *Astronomy*

### *Fred Hoyle*
CALIFORNIA INSTITUTE OF TECHNOLOGY

W. H. FREEMAN AND COMPANY
*San Francisco*

**Library of Congress Cataloging in Publication Data**

Hoyle, Sir Fred.
    Highlights in astronomy.

    Includes index.
    1.  Astronomy.    I.  Title.
QB45.H89       520         75-1300
ISBN 0-7167-0355-6
ISBN 0-7167-0354-8 pbk.

Printed in the United States of America

9 8 7 6 5 4 3 2 1

# Preface

In the very troubled times in which we are living today, astronomy offers a welcome new perspective on a scheme of things much larger than ourselves. This perspective, being based on meticulous observation and firm reasoning, is not a form of escapism, but a discipline founded in hard fact; yet it offers the kind of horizon that many people are seeking today.

In this book, within the compass of less than 200 pages, I have tried to present something of the grandeur and the precision of the things we know about. To the scientist, the wonderment of our world is that the things we know about are almost surely a mere fraction of the things which still lie beyond the range of our understanding. This is exactly why the scientist is driven ever onward, always seeking to rend aside the obscuring veil a little further, so that new aspects of the universe become known to us all. Each new discovery is a precious addition to the hard-won store of knowledge possessed by the human species.

I would like to thank my wife, Barbara, for her invaluable help in organizing the structure of this book.

*Pasadena, California*                                    *Fred Hoyle*
*December 1974*

# Contents

1 The Earth 1

2 The Planets and the Sun 25

3 The Planets Themselves 37

4 The Sun 65

5 Comets and Other Forms of Debris 81

6 Stars 95

7 Life in the Universe 127

8 Galaxies and the Universe 139

Questions and Topics for Discussion 171

Index 177

# 1

## *The Earth*

# 1

## *The Earth*

Clearly, to be able to see the whole Earth, one must see it from the space outside the Earth. Figure 1.1 shows the launch of Apollo 17. This was the beginning of a journey to the Moon. Figure 1.2 shows the full Moon.

*Apollo* was the code name for a series of missions to the Moon organized by the National Aeronautics and Space Administration (NASA). The first actual landing on the Moon came on the eleventh of these missions, Apollo 11, but the first picture of the Earth was taken earlier, by the crew of Apollo 8. This picture, seen in Figure 1.3, shows the Earth rising over the Moon. A more recent picture of the Earth, taken on the flight of Apollo 17, is shown in Figure 1.4.

Why can the whole round ball of the Earth be

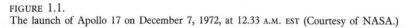

FIGURE 1.1.
The launch of Apollo 17 on December 7, 1972, at 12.33 A.M. EST (Courtesy of NASA.)

FIGURE 1.2.
The full Moon, photographed on the Apollo 11 mission.
This is the side of the Moon seen from the Earth.
(Courtesy of NASA.)

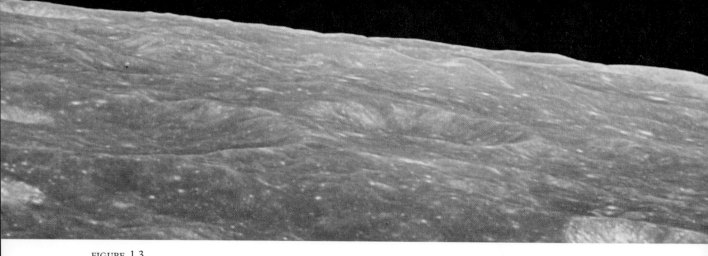

FIGURE 1.3.
Earth-rise over the Moon, from the Apollo 8 mission. When the Earth is seen full from the Moon, it is some seventy-five times brighter than the full Moon seen from the Earth. (Courtesy of NASA.)

FIGURE 1.4.
Apollo 17: A view of the Earth on the journey toward the Moon.
(Courtesy of NASA.)

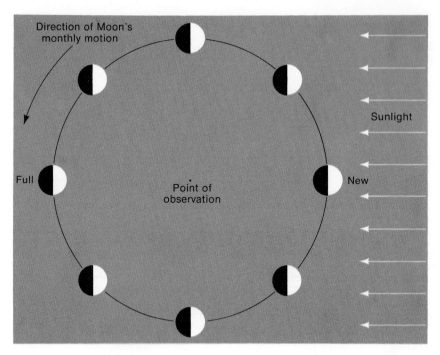

FIGURE 1.5.
As the Moon goes around the Earth, we see different aspects of the sunlit half of the Moon.

seen in Figure 1.3 but not in Figure 1.4? For the same reason that the Moon is most often seen as a crescent or as an incomplete ball, a reason that can be understood from Figure 1.5. The Moon and the Earth shine because they reflect sunlight. On the other hand, the stars, which can be seen in the sky on any clear night, all shine because they generate light of their own. Indeed, stars are bodies like the Sun.

It is interesting to take a close look at Figure 1.4. The vast land mass of Africa can be seen at the upper left, connected to Arabia at top center. At the upper right, a spiral swirl of cloud is a storm in the Indian Ocean. Clouds show white and they tend to be curved in spirals, an effect caused by the

spin of the Earth. As we know, the Earth spins around once every 24 hours with respect to the Sun, and it is this spinning which produces the sequence of night and day. Figure 1.6 shows the spin of the Earth in relation to the Sun for a moment when it is dawn across the American continent, on a day in the year close to midsummer.

The white mass at the bottom of Figure 1.4 is not cloud, however, but ice. This is the Antarctic, the scene of Captain Scott's heroic walk to the South Pole. When one looks carefully, it is not hard to see where the ice ends. At its boundary, the Antarctic ice meets the dark, forbidding southern oceans, which navigators of the eighteenth century, particularly Cook, tried to penetrate. It was in these

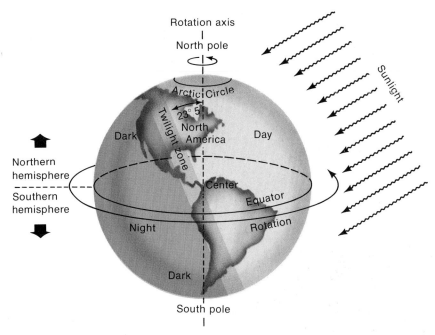

FIGURE 1.6.
The Earth rotates with respect to the Sun once in 24 hours, which causes the sequence of night and day. This figure shows a moment when there is dawn across the American continent.

waters that the whaling industry of Herman Melville's *Moby Dick* became established in the nineteenth century.

It is somewhat ironic that it should have fallen to the lot of such an earth-bound animal as Man to have obtained the picture of Figure 1.4. Our daily vision of the Earth is limited to the office and the street, or at most to a few square miles of open country. The creature who habitually sees the Earth most in the spirit of Figure 1.4 is perhaps the albatross, for the Earth as a whole is the territory of the albatross. Unlike the man-made jet airplane, far-flying birds like the albatross, and like the swift and shearwater, use very little energy of their own. They gain lift and speed from upward air currents and from wind. The energy which generates these atmospheric motions come from sunlight. So birds travel in effect on sunlight.

The astronauts who landed on the Moon, in the kind of terrain shown in Figure 1.7, were exhilarated to find their weight to be much less than it is on the Earth. A man who weighs 180 pounds on the Earth weighs only about 30 pounds on the Moon. It is interesting to consider what would have happened on the Earth if the downward pull of gravity, which causes our weight, had happened to be as weak here as it is on the Moon.

The idea that all the plants and animals on the Earth have evolved slowly during a long period of time from a simple beginning became widespread

FIGURE 1.7.
Apollo 17: Rover vehicle on the desolate lunar surface.
(Courtesy of NASA.)

FIGURE 1.8. (*facing page*)
How evolution is believed to have occurred
on the Earth. Geological ages are given
along the bottom. Notice the connections
of one sequence of evolution with another,
denoted by dotted lines. (From Shelton,
*Geology Illustrated,* W. H. Freeman and
Company. Copyright © 1966.)

in the nineteenth century, although it seems to have been proposed first by Robert Hooke, a contemporary of Isaac Newton (1643–1727). Biologists for many decades have been much concerned with discovering the details of how this evolution actually took place. The information used in this work comes mostly from fossils. Fossils result when plants and animals become trapped in rock slides or quicksands, or become washed underground and covered by debris. Sometimes the forms of such plants and animals become communicated to surrounding material, as when an animal is entombed in clay. Then the clay sets, and eventually becomes hard rock. So plant and animal forms can some-

times be found when rocks are broken open. Since geologists can supply approximate dates for when the rocks in question were formed, it becomes possible to determine when particular plants and animals flourished on the Earth. It is in this way that the flourishing of the dinosaurs some 100 million years ago was discovered. Figure 1.8 gives details of the biological history of the Earth during the last 500 million years, a period equal to about ten million human generations.

The details of Figure 1.8 depend on the physical properties of the Earth. If the physical properties had been different, for example, if gravity had been less, as it is on the Moon, Figure 1.8 would have

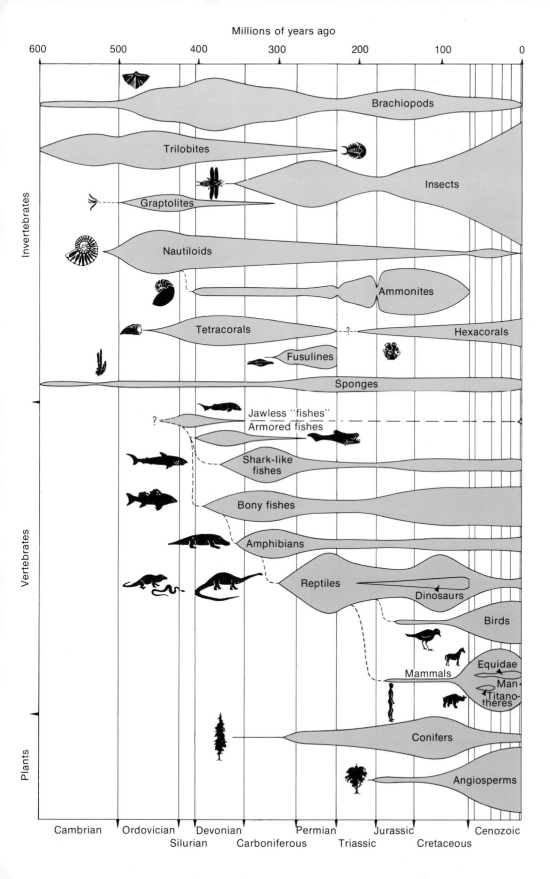

Millions of years ago

Invertebrates

Brachiopods
Trilobites
Insects
Graptolites
Nautiloids
Ammonites
Tetracorals
Hexacorals
Fusulines
Sponges

Vertebrates

Jawless "fishes"
Armored fishes
Shark-like fishes
Bony fishes
Amphibians
Reptiles
Dinosaurs
Birds
Equidae
Man
Titano-theres
Mammals

Plants

Conifers
Angiosperms

Cambrian    Ordovician    Devonian    Permian    Jurassic    Cenozoic
Silurian    Carboniferous    Triassic    Cretaceous

9

been different. The connections indicated in Figure 1.8 by dotted lines, show that the mammals—the class of animal to which the human belongs—have an ancestry that passed through bird-like creatures. Birds are related to us in a way that flies, wasps, and bees are not. In our emotional attitude to birds and insects, we recognize this difference instinctively.

If gravity were less on the Earth, there is no doubt that birds would be better placed in relation to the whole of evolution. As things are at present, birds are nearly too heavy to fly at all. They do so only by paring their weight to a minimum: they have hollow bones and brains of minimal capacity, and for long-ranging birds even the ability to come freely down from the air onto land is denied. The swift, for example, must always nest in such a position that it can take to the air by falling out of the nest. If gravity were less, it would be possible for far-flying birds to be equipped with more armament for attacking land-based creatures. Instead of carrying off a lamb, an eagle might carry off a man. Still more formidable, it would be possible for birds to be endowed with thinking brains. In such circumstances, it would be unlikely that land-based creatures like Man could have become established at all.

These are not just fanciful speculations of no relevance to the world. The situation we have just considered could well arise on another planet moving around some other star.

Although the development of plants and animals shown in Figure 1.8 may seem to have occupied a very long span of time, some 500 million years, the Earth itself is nearly ten times older than this. The age of the Earth has been found to be about

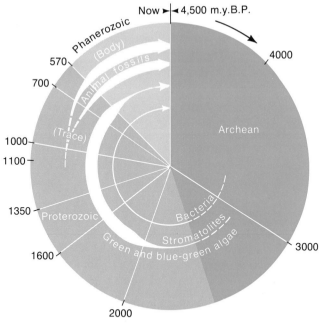

FIGURE 1.9.
The early forms of life were bacteria
and the blue-green algae.

4,500 million years; if we think of a human generation as occupying 45 years, the Earth has been
moving year by year around the Sun for as long as
100 million human generations.

A way of representing this great span of time is
shown in Figure 1.9. Imagine the hand of a clock
starting in the 12 o'clock position, and imagine it
making a complete circuit, but taking 4,500 million
years to do so. That is to say, the hand starts 4,500
million years back into the past, at the time the
Earth was formed. The numbers marked on the
outer circle of Figure 1.9 refer to moments of time

in the past—4,000 means 4,000 million years ago,
3,000 means 3,000 million years ago, and so on.
Also shown in Figure 1.9 are the broad forms of
life that have existed in the past on the Earth. The
detailed forms of Figure 1.8 all belong to the last
sector, starting at the number 570 (570 million
years ago) and continuing through to the present
day.

Life has existed on the Earth for about 3,500
million years, but at first only in the simple forms
of bacteria. Next came the green and blue-green
algae, going back nearly 3,000 million years. These

early forms have the characteristic of being able to live at quite high temperatures. Some forms of bacteria can indeed survive at the temperature of boiling water. The algae grow in profusion in volcanic hot springs like those of Yellowstone National Park. The fact that the early life forms were of this high-temperature kind suggests either that life started in hot springs, or that the whole Earth was very much warmer than it is at present.

Although it may be hard at times to understand the complicated details of the problems on which scientists are engaged, it is not hard to understand the broad features of the biggest problems. The origin of life is one of the big problems. Many scientists are at present trying to understand how life may have started from only quite simple chemical substances, like water, ammonia, and the gas carbon dioxide ("dry ice" when frozen). This problem has relevance to whether life may also have started on planets like the Earth that are moving around other stars similar to the Sun.

The familiar shapes of the Earth's continents undergo major changes when time intervals as long as a few hundred million years are considered. Figure 1.10 shows how the coastline of West Africa matches the eastern coastline of South America. Not only do the shapes match, but particular kinds of rock are found to fit together when the two coastlines are brought together as they are in Figure 1.10, in which the fit of the rocks is shown by the dark areas.

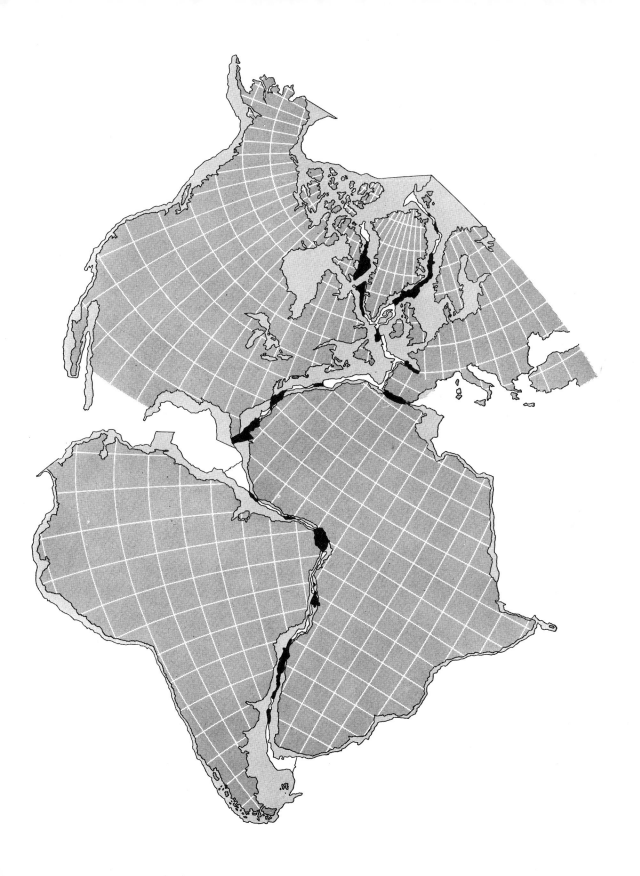

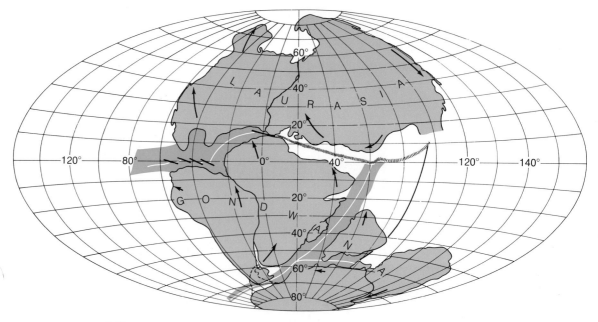

FIGURE 1.11.
In a similar way, the east coast of the United States can be fitted to Africa, and Europe to both Africa and the United States. This association of the continents is believed to have existed about 200 million years ago. (After R. Dietz and J. Holden, "The Breakup of Pangaea." Copyright © 1970 Scientific American, Inc. All rights reserved.)

The explanation given for this situation is simple but remarkable, namely, that Africa and South America were once joined together. Indeed, other sections of the continental land masses were once joined together, in the pattern of Figure 1.11, which shows the continents as they were some two or three hundred million years ago—about half-way through the final sector of the "clock" of Figure 1.9.

The continents are made out of rock that is some 25 per cent less heavy, volume for volume, than the rocks on which the continents lie. The continental rocks are also less heavy than the rocks of the ocean floors. This causes the continents to "ride high." The lighter rock floats in the heavier rock. Figure

1.12 shows how the western United States rides high above the floor of the Pacific Ocean.

Although the evidence of Figure 1.10 has been available for a long time, scientists were reluctant to believe the obvious implication—that Africa and S. America were once joined together—because enormous forces would be needed to tear two such land masses apart. Nowadays the forces are thought to arise from the operation of a kind of nuclear engine within the Earth itself. The energy source for the engine comes largely from the element uranium, which is the same element used to supply the energy output of a man-made nuclear reactor. Indeed, the physical process of energy release is the same, namely, break-up of uranium atoms into

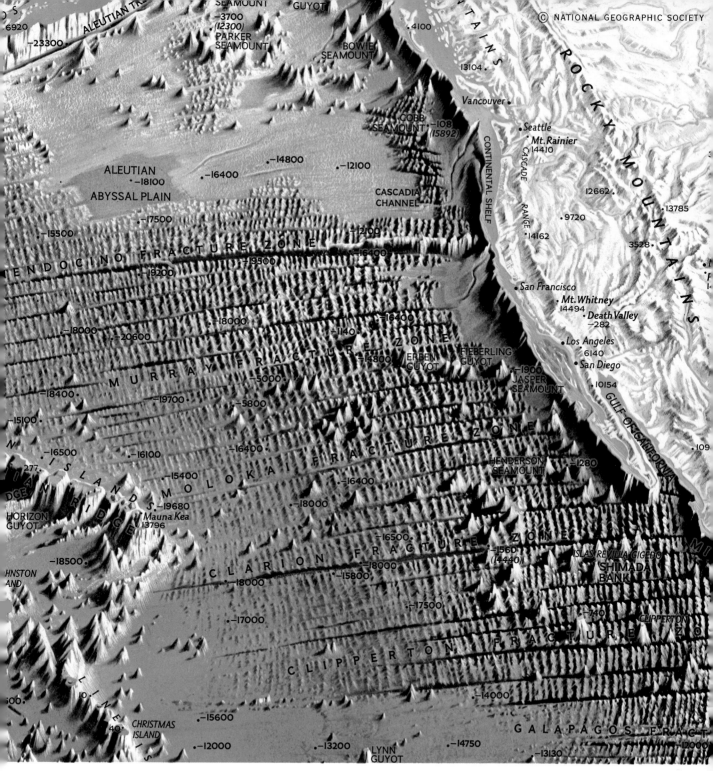

FIGURE 1.12.
A part of the Pacific Ocean floor.
(Courtesy of the National Geographic Society. Copyright © 1968.)

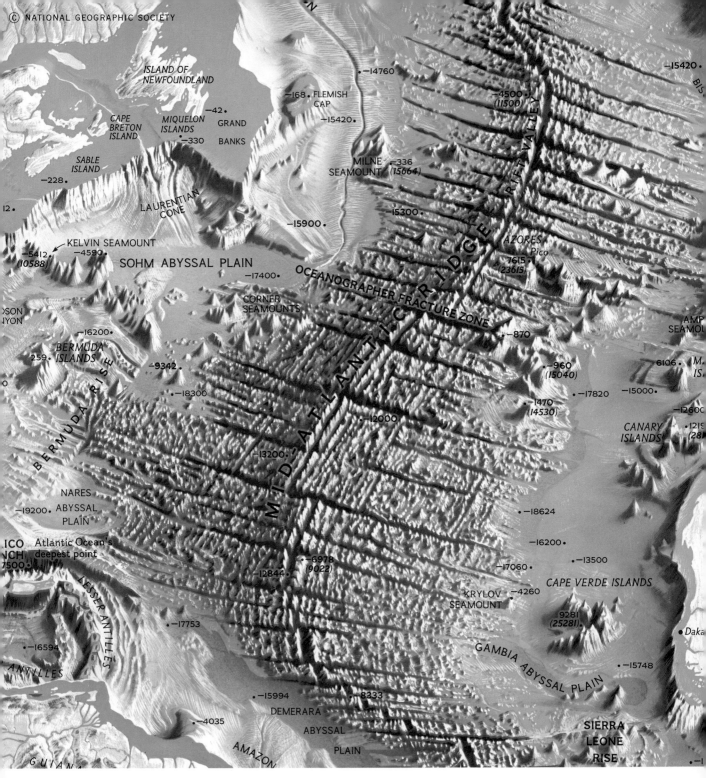

FIGURE 1.13.
A part of the Atlantic Ocean floor.
(Courtesy of the National Geographic Society. Copyright © 1968.)

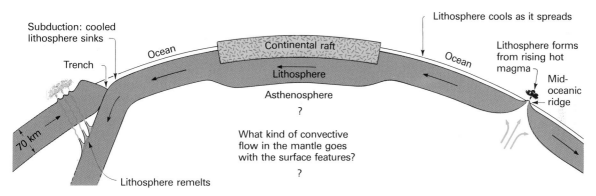

Subduction: cooled
lithosphere sinks

Trench

Ocean

Continental raft

Lithosphere

Asthenosphere

?

What kind of convective
flow in the mantle goes
with the surface features?

?

70 km

Lithosphere remelts

Lithosphere cools as it spreads

Ocean

Lithosphere forms
from rising hot
magma

Mid-
oceanic
ridge

FIGURE 1.14.
The rocks of the Earth's crust move in a system of plates, producing ridges when material emerges
from the interior, and causing deep troughs when material goes back to the interior. The lighter rocks
on the continents "float" in this system of moving plates. (After J. Dewey, "Plate Tectonics." Copyright
© 1972 by Scientific American, Inc. All rights reserved.)

two pieces, a process known as fission, which
generates heat.

What has happened in the Earth is that so much
heat has by now been released that an engine has
been set in operation, an engine in which the
Earth's outer crust has been set in motion. The
motion is rather complicated: the crust is divided
into pieces, "plates," as they are called, which all
move with respect to each other. At some places
the rocks of a plate emerge from the Earth's interior, as they do along the mid-Atlantic ridge shown
in Figure 1.13. At other places the rocks of one
plate may dip below those of another plate, or
below the rocks of a continent. The general idea of
this plate movement is shown in Figure 1.14.

The continent of Africa rides on a plate, as does
the connecting part of Asia. These two plates are
different, and their motions are causing them to
press against each other. It is this pressing together
of plates which has caused a vast line of mountains
to be pushed up, starting in the west of the Alps,
passing through Turkey and the Caucasus,
and eventually reaching Afghanistan and the
Himalayas.

Sometimes a plate presses against a continental
mass, as it is doing along the western coastline of
the Americas, causing the mountain ranges of the
Rockies and Andes to be formed. In still other
places, the rocks of two plates dip down together,
producing the deeps of the ocean floor. The whole

complex of all these plates, over the whole surface of the Earth, is shown in Figure 1.15. It is their motion that causes volcanoes and earthquakes: volcanoes result from the heat generated by friction when rocks grind against each other, and earthquakes result from the actual motions themselves. These dramatic phenomena occur where one plate abuts another. In contrast, the central regions of the plates, far removed from their edges, tend to be rather quiescent. This is why some places on the Earth are violent and some peaceful; which one a place will be depends on where it is with respect to the Earth's system of plates.

We tend to think of earthquakes and of the outbursts of volcanoes as disasters, and we tend to regard mountain ranges as barren regions. So we might think that the plate motions are not an advantage to human life. Yet on a broader view, we would be quite incorrect. Without this continuous turning over of the Earth's crust, there would probably be no mineral deposits on the Earth's surface. Suppose there were a range of 20,000 foot mountains in the center of Australia. It would bring down vast quantities of snow, and large rivers would flow out of such a mountainous area, which would provide much needed water throughout what is now the great Australian desert. So we can see that, without the system of plate movements, the surface of the Earth would almost certainly be far more inhospitable to man. Indeed, the gradual

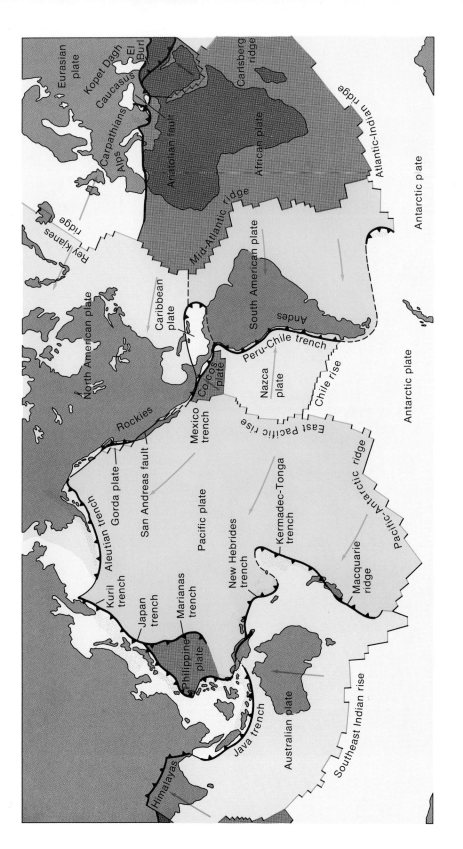

FIGURE 1.15.
Showing how the surface of the Earth is broken up into plates.
(Allen J. Dewey "Plate Tectonics." Copyright © 1972 by
Scientific American, Inc. All rights reserved.)

———— Subduction zone
———— Transform
———— Ridge axis

–––––– Uncertain plate boundary
→ Direction of plate motion
Areas of deep-focus earthquakes

FIGURE  1.16.
Apollo 13: Another view of the Earth.
(Courtesy of NASA.)

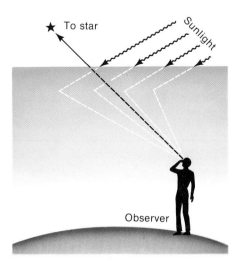

FIGURE 1.17.
Blue light is scattered more strongly than red light by the particles of which air is composed, causing the sky to appear blue.

wearing down of the continents might well smooth out the Earth so much that, instead of the continents sticking out of the ocean, water would entirely cover the whole surface of the Earth. The nuclear engine inside the Earth thus does much more than provide us with interesting scenery; it provides the environment for life as we know it.

In Figure 1.16 we have another picture of the Earth, taken on the homeward flight of Apollo 13. All these pictures have a blue glow around the edge, which comes from the Earth's atmosphere. This blue color arises in the same way as the familiar blue of the clear daylight sky. How this happens can be understood from noticing that ordinary sunlight consists of light of many colors, as we can see in the phenomenon of the rainbow. What happens in a rainbow is that small drops of water in the air break up the sunlight into its constituent colors. The molecules of the air are much smaller particles; these also affect light of different colors differently. They scatter blue light more strongly than other colors, thereby causing blue light, that would otherwise never have reached us, to be redirected toward us. This scattering effect is seen in Figure 1.17.

Dust particles in the air, which may be raised by

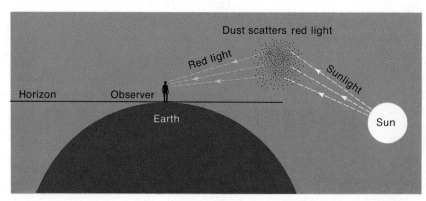

FIGURE 1.18.
Dust particles suspended in the air scatter red light from the setting Sun more strongly than blue light.

windstorms or injected into the air from factory chimneys, also scatter light, but they behave in the opposite way from the much smaller particles of the air: they scatter red light more strongly than blue light. This scattering of red light by dust particles often has the effect shown in Figure 1.18, and so creates the rich sunset in Figure 1.19.

FIGURE 1.19.
Dust particles and droplets cause the redness of a sunset.
(Courtesy of Robert Ishi.)

# 2

## The Planets and
## the Sun

# 2

## *The Planets and the Sun*

The Earth goes around the Sun once in a year—in fact, this is exactly what we mean by the year, the length of time the Earth needs to make a complete circuit of the Sun. The path of the Earth is nearly a circle, as it has been drawn to be in Figure 2.1. This path, or "orbit" as it is usually called, is very much larger than either the Earth or the Sun. So to get a realistic idea of scale, one should think of the Sun in Figure 2.1 as a small dot, and of the Earth as an even smaller dot.

Yet it is useful to redraw Figure 2.1 with the sizes of the Sun and the Earth grossly exaggerated, as in Figure 2.2. This second figure is also wrong in showing the Earth to be about the same size as the Sun, whereas the diameter of the Sun is in fact about a hundred times larger than the diameter of the Earth. The point of these deliberate mistakes in Figure 2.2 is to bring out the relation of the daily spin of the Earth to the yearly motion of the Earth around the Sun. The Earth spins around an axis

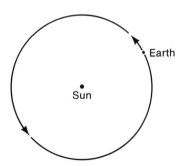

FIGURE 2.1.
In comparison with the scale of its orbit around the Sun, the Earth is only a speck.

which joins the north and south poles, and this axis always points in the same direction as the Earth pursues its orbit around the Sun. It will be seen that the axis is not perpendicular to the plane of the orbit, but is inclined to such a perpendicular direction by about 23.5° (a right angle has 90° and a complete turn has 360°, as shown in Figure 2.3).

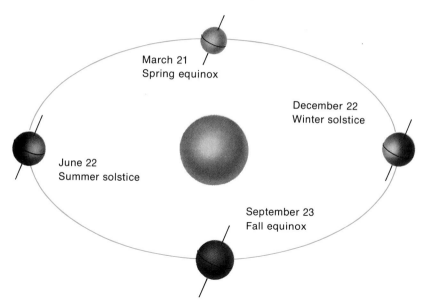

FIGURE 2.2.
The sizes of the Sun and the Earth are both grossly exaggerated in this figure. The Earth moves around the Sun in an orbit that is nearly a circle, and the axis of the Earth's daily rotation maintains a fixed direction. The varying orientation of the Sun with respect to the two hemispheres of the Earth is responsible for the seasons. (From J. Brandt and S. Maran *New Horizons in Astronomy*. W. H. Freeman and Company. Copyright © 1972.)

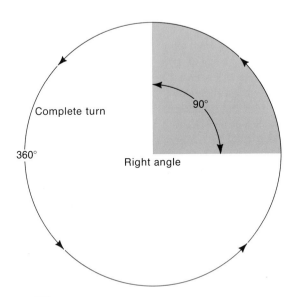

FIGURE 2.3.
There are 90 degrees (90°) in a right angle, and 360° in a full turn.

If it had happened that, instead of this tilt of 23.5°, the axis of the Earth's rotation had been exactly perpendicular to the plane of the orbit, then every day would have had 12 hours turned toward the Sun and 12 hours turned away from the Sun, and the situation in the northern hemisphere of the Earth would always have been the same as in the southern hemisphere. Because of the tilt, however, the situation changes all through the year. At point A of Figure 2.2 the northern hemisphere leans most toward the Sun, giving more than 12 hours of daylight in the northern hemisphere and less than 12 hours in the southern hemisphere. At the point B on the opposite side of the orbit the situation is reversed, with the southern hemisphere leaning most toward the Sun, and thus with the southern hemisphere receiving more than 12 hours of daylight. In fact, point A corresponds to midsummer and point B to midwinter in the northern hemisphere, and vice versa for the southern hemisphere. So it is the tilt of the Earth's axis of rotation which causes the seasons of the year.

There are two lengths of time which are of practical importance to us. We divide our activities into days, because this gives a convenient relationship to the hours of daylight and of darkness. On the other hand, the year is also important because

the year gives a convenient relation to the seasons. A complication arises, however, since there is no inherent relation between these two different ways of reckoning time. There is no astronomical reason why there should be an exact number of days in the year, and in fact there is not an exact number. As near as is known, there are 365.242194 . . . days in the year, and the decimal here probably never terminates.

This raises the practical problem of how we can continue to divide time into days and yet prevent the year from getting "out of step." For example, if we decided always to count 365 days in the calendar year, we should be in error compared to the true astronomical year by 0.242194 . . . days every time the Earth made a circuit around the Sun. After only ten circuits our artificial calendar would be out of step with the real astronomical year by 2.42194 . . . days, and after a hundred times around the Sun by 24.2194 . . . days. In a century our calendar would be very seriously in error. Evidently a better method for constructing a calendar is needed.

The method introduced by Julius Caesar in the year 46 B.C. consisted in letting every fourth year have 366 days; indeed, we are all familiar with this "leap year" system in which an extra day is added to the month of February. This Julian system, as it is called, makes the average calendar year equal to 365 days, which is still not quite the same as the actual year. The error in the Julian system was still .0078 . . . days for every circuit of the Sun by the Earth, so that by the time Caesar's calendar had been in operation from 46 B.C. to A.D. 1500, an error of about 12 days had accumulated. As a result, a more complicated prescription for the leap days was proposed in 1582 by Pope Gregory. This Gregorian system departed from Caesar's calendar by leaving out three leap years during every four centuries. For example, the year 1900 would have been a leap year according to the Julian calendar but not according to the Gregorian calendar.

The Gregorian calendar was immediately adopted in all Catholic countries, but was not adopted in Protestant England until the eighteenth century. It is the system currently in use, the system whereby the calendars we use in our annual diaries are constructed. Yet Pope Gregory's prescription, like that of Caesar before him, is not completely accurate. It gives an average calendar year of 365.2425 days, still an error of about .0003 days for every circuit of the Sun by the Earth. If the human species survives for another 40,000 years, and if the Gregorian calendar continues in use, the calendar year will again become out of step with the true year by as much as 12 days. There is *no* way to produce a *regular* prescription of the kind given by Caesar and by Pope Gregory that will not eventually run into serious error.

By an *irregular* prescription, one can do very much better, however. The rules are simply these:

If the calendar year is not out of step with the true year by more than 1 day, the calendar year has 365 days. But if the calendar year is out of step by more than 1 day, the calendar year has 366 days.

Under these easily understood rules the error can never exceed a day, however long they operate (provided of course that the day and the true year do not change for genuine astronomical reasons). If this system were adopted, the leap years of 366 days could be determined long in advance, so no practical inconvenience would arise. The leap years would be found not to fall in quite a regular sequence, and it would be precisely in its irregularity that this system would be superior to the calendars of Caesar and of Pope Gregory.

Besides the Earth there are eight other planets, all of which move in orbits around the Sun. These orbits are shown in Figure 2.4. Because their orbits all have different scales, other planets take different

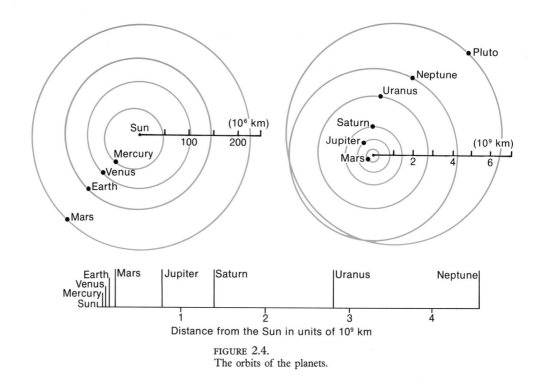

FIGURE 2.4.
The orbits of the planets.

TABLE 2.1.
*Planetary orbits*

| Planet | Average distance[a] of planet from the Sun | Time required for complete circuit (years) |
|---|---|---|
| Mercury | 0.387 | 0.241 |
| Venus | 0.723 | 0.615 |
| Earth | 1.000 | 1.000 |
| Mars | 1.524 | 1.881 |
| Jupiter | 5.203 | 11.86 |
| Saturn | 9.540 | 29.46 |
| Uranus | 19.18 | 84.01 |
| Neptune | 30.07 | 164.8 |
| Pluto | 39.44 | 248.4 |

a. Here the average distance of the Earth from the Sun is used as the unit of measurement; this distance is known as the *astronomical unit* (A.U.).

lengths of time to make a complete circuit of the Sun. Table 2.1 gives the average distance of each planet from the Sun, and also the number of Earth years each planet requires for a complete circuit.

The unit of distance used in Table 2.1 is the Earth's own average distance from the Sun, which is known quite accurately, and is 149.6 million kilometers. So to get the actual average distances from the Sun to the other planets, one simply multiplies 149.6 million kilometers by the numbers which appear in the middle column of Table 2.1.

From Figure 2.4 it might seem as if the orbits of all the planets lie in the same plane. Although this is nearly true, it is not strictly true. The planes of the different orbits are actually inclined to each other by small angles of a few degrees. These small inclinations complicate the paths which the planets

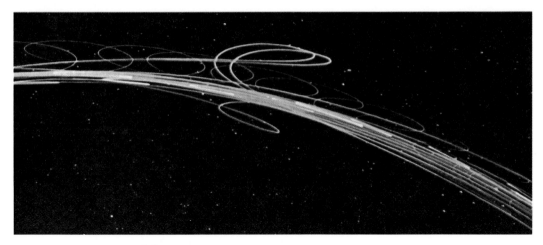

FIGURE 2.5.
The apparent motions of the planets on the sky, simulated in a planetarium.
(Courtesy of Planetarium Munich.)

follow in the sky. Figure 2.5 is a reconstruction of how the different planets appear to move when viewed from the Earth.

If one started by knowing Figure 2.4 and the numbers set out in Table 2.1, it would not be hard to work toward Figure 2.5. It is much harder, however, to work from Figure 2.5 to Figure 2.4, and this was what the early astronomers had to do, since Figure 2.4 was not known to begin with. The first step in disentangling Figure 2.5 comes from realizing that much of its complexity arises from the Earth's own motion, the orbital motion of Figure 2.1.

The man who first understood that this was so, and who managed to win over other astronomers to his point of view, was Copernicus (1473-1543); Figure 2.6 is believed to be his self-portrait. The diagram which Copernicus himself drew of the planetary orbits is shown in Figure 2.7, which is a reproduction of a page from the original manuscript of his book *De Revolutionibus Orbium Caelestium*. This diagram does not show the three outermost planets, which were not known to Copernicus. Uranus was discovered in 1781, Neptune in 1846, and Pluto in 1930.

FIGURE 2.6.
Nicolaus Copernicus, 1473-1543.
(Courtesy of Muzeum w Toruniu.)

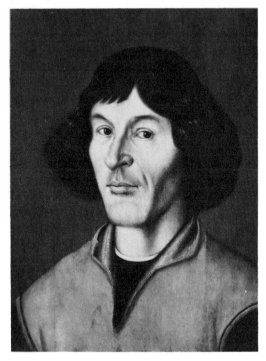

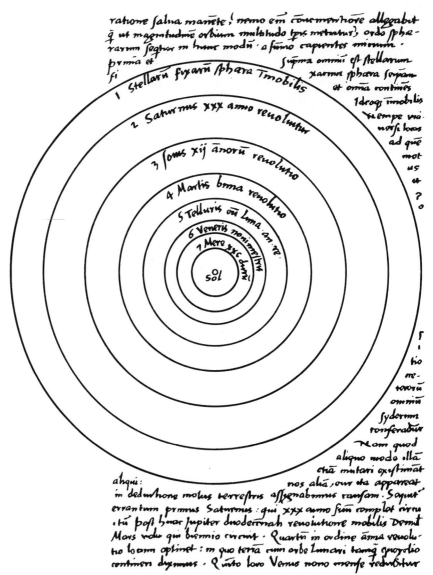

FIGURE 2.7.
A page from Copernicus'
original manuscript.

FIGURE 2.8.
Johann Kepler 1571-1630.
(Culver Pictures.)

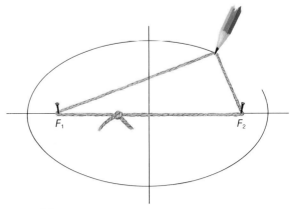

FIGURE 2.9.
How to draw an ellipse. The pins at $F_1$ and $F_2$ mark the two foci of the ellipse. (From A. Baez, *The New College Physics*. W. H. Freeman and Company. Copyright © 1967.)

It will be seen that Copernicus' drawing represents the orbits as being circles centered at the Sun, whereas Figure 2.4 shows the orbits, especially of Mercury and Mars, are not centered on the Sun. Copernicus understood this perfectly well, and indeed he spent his whole life seeking a geometrical construction to describe correctly the departure of each orbit from a simple circular form. In this he was only partially successful, and it was left to Kepler (1571-1630; Figure 2.8) to first state the correct form for the orbits: they are all ellipses.

A simple construction for an ellipse is shown in Figure 2.9, the two points $F_1$ and $F_2$ where the string is attached in this figure being known as the "foci" of the ellipse. By varying the distance between $F_1$ and $F_2$, and by varying the length of the string, every possible ellipse can be drawn. The planetary orbits are all of a kind in which the length of the string is long compared to the distance from $F_1$ and $F_2$—that is, they are all *nearly* circles. Kepler discovered another remarkable fact. Each planet has an orbit which is an ellipse, and the Sun always lies at one of the foci, either at $F_1$ or $F_2$ in the construction of Figure 2.9.

Once the ellipse forms of the planetary orbits became known, men began to ask the question *why*. Taking the simplest problem to begin with, if one goes back to Figure 2.1 for the orbit of the Earth (thereby neglecting the slight difference between the true elliptic orbit and a circle), what is it that keeps the Earth moving endlessly round and round the Sun? Why doesn't the Earth simply fly off at a tangent, leaving the Sun altogether? We can come to grips with this problem by means of an analogy,

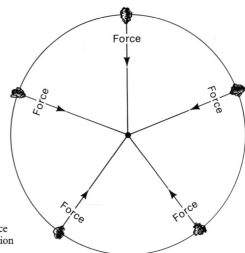

FIGURE 2.10.
To keep a stone whirling around a circle, a force must be applied to the stone, always in a direction toward the center of the circle.

in which a weight at the end of a piece of string is whirled around in a circle (Figure 2.10). This is something which everybody can try for themselves. It immediately becomes clear, as soon as a trial is made, that a pull must be exerted on the string in a direction from the weight toward the center of the circle. Without this pull the weight certainly does fly off at a tangent—a fact made use of in the weapon known as a sling. This suggests the answer we are seeking. The Earth goes round and round in Figure 2.1 because it is subjected to a pull in a direction toward the Sun. Since there is no string in Figure 2.1, the pull must reach across the space between the Sun and the Earth. This pull became known as the force of gravitation. It was quickly recognized to be the same force as the ordinary gravity which holds us all down onto the surface of the Earth. We do not fall off the Earth, because the Earth exerts a downward pull on us, and the Earth itself does not fly away from the Sun because of the pull which the Sun exerts on the Earth.

But would this idea of a force directed from a planet toward the Sun, so useful if the orbits of planets were circles, be capable of explaining the actual elliptic forms of the orbits? This problem was not so easy. It was one with which scientists wrestled during the last quarter of the seventeenth century, a problem eventually solved by Isaac Newton (1643-1727). The answer turned out to be triumphantly affirmative: the same gravitation which worked for a falling stone or a falling apple on the Earth, and which worked for circular orbits, also worked for elliptic orbits.

The solution to this critical problem, a major landmark in the history of science, was achieved by Newton because, unlike most scientists, Newton was not just a user of known mathematics. He was capable of inventing whole new fields of mathematics should the occasion warrant it. In later life he invented a new branch of mathematics simply to solve an exercise. The exercise was set as a challenge to the mathematicians of Europe by Gottfried Wilhelm Leibniz (1646-1716) and Johann Bernoulli (1667-1748). By this time, the

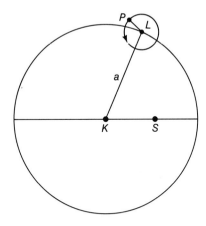

FIGURE 2.11.
In the construction of Copernicus, the line *KL* turns at
a uniform rate. The line from *L* to the planet at
P turns at twice this rate, the distance *L* to *P* being
1/3 of the distance from *K* to the Sun at *S*.

year 1697, Newton had retired from active work as a scientist. He had come to disdain science as a "philosophical conceit," and he disdained to attempt the problem set by Leibniz and Bernoulli. When after six months no one had solved the problem, Bernoulli renewed the challenge, making a special point of communicating it to Newton. This direct approach left Newton with little alternative but to accept. His niece records tersely in her diary for January 29, 1697, "Bernoulli sent problem. I.N. home at 4 P.M. Finished it by 4 A.M." True to his peculiar personality, Newton insisted that the solution be published anonymously. When Bernoulli saw it, he is said to have exclaimed, "Ah! I recognize the lion by his paw!"

It might seem strange that it should have taken nearly two thousand years to arrive at Kepler's discovery of the elliptic form of the planetary orbits. In retrospect, we can see that this discovery was made much more difficult by the ancient belief that the orbits must be compounded from circles. This belief came down from the Greeks,

who regarded the circle as the most perfectly formed curve. Since the heavens were thought by both ancient and medieval astronomers to be a representation of perfection, it followed therefore that the planets could only move in circles or in combinations of circles. Everybody felt that the problem *had* to be soluble in these terms. It came as a most profound surprise to Kepler when he found this was not the case. Even so original and far-sighted an astronomer as Copernicus never sought in his life-long attack on the problem to consider anything but combinations of circles. Figure 2.11 is typical of the constructions tried by Copernicus.

Writers on scientific method usually tell us that scientific discoveries are made "inferentially," that is to say, from putting together many facts. But this is far from being correct. The facts by themselves are never sufficient to lead unequivocally to the really profound discoveries. Facts are always analyzed in terms of the prejudices of the investigator. The prejudices are of a deep kind, relating to our views on how the universe "must" be constructed.

Just as each planet is subject to a gravitational pull from the Sun, so the planets exert gravitational forces on each other. These extra forces are small compared with the effect of the Sun, but they do have minor influences on the planetary orbits, causing them to depart very slightly not just from circles but even from Kepler's ellipses. If one insists on considering all these small effects, the planetary orbits are indeed exceedingly complex, without any simple geometrical description at all. Had the Greeks been told of this lack of all ultimate geometrical simplicity, they would undoubtedly have been profoundly shocked. Their prejudices on how the world "must" be constructed would have collapsed.

Nowadays our prejudices are more sophisticated, but they are everpresent nonetheless. We still maintain the Greek idea of perfection, of aesthetic elegance, not in the phenomena themselves, but *in the underlying laws which govern the phenomena.* We are not worried today that the forms of the planetary orbits seem almost capriciously complex.

We pass this by with the comment that it all depends "on how the system was started off in the first place." For planets moving around another star, we expect different orbits. There is no explanation for the intimate details of our own system except in the particularities of how it began. We in turn would be deeply shocked if we were to find the laws governing the planetary motions to be of a capricious nature. It was just because Einstein felt Newton's theory of gravitation contained capricious elements that he was led to a new and different theory of gravitation, one that turned out to be triumphantly vindicated.

When I was young, not many years into research, Paul Dirac once said to me, "Hoyle, you are much too empirical. Look more closely at the mathematical structure of what you are doing." When I asked what it was I should look for, Dirac answered, "You have to learn to recognize what is beautiful." More closely than anything else which can be said in words, this describes the deeply felt belief that today inspires scientists to attempt to delve farther into the structure of the universe.

# 3

## *The Planets Themselves*

# 3

## *The Planets Themselves*

Scientists frequently make use of a concept known as "mass." We have an instinctive idea that the mass of a body has something to do with its bulk and with its weight, but these notions are not precise enough. A bushel of feathers and a bushel of lead may have the same size—the same volume— but they certainly do not have the same mass. In some degree weight is a better indicator of mass. If we weigh two bodies under similar circumstances, and one body is found to have twice the weight of the other, then is it correct to say that the mass of the heavier body is twice the mass of the lighter body. This way of comparing masses, by means of their weights, is useful when we want to consider the different kinds of atoms out of which all material is constructed.

Suppose we arrange to weigh the *same number* of atoms whatever the kind. Then from the measured weights we will know how the mass of one kind of

atom compares with the mass of another kind. In this way it is known that the atom of hydrogen has the least mass of all the different kinds of atoms. It turns out that the carbon atom, for example, has about 12 times the mass of the hydrogen atom, oxygen has 16 times, aluminum 27 times, iron 56 times, and so on.

The mass of a large body is simply the mass of its constituent atoms. Imagine the body divided into atoms, and proceed to separate the atoms into their various kinds. Count the number of atoms of each kind, and then take care to allow for the fact that different atoms have different masses. Work in terms of hydrogen, the atom of least mass. Thus for hydrogen count 1, but for each carbon atom count 12, because each carbon atom has twelve times the mass of a hydrogen atom. For each oxygen atom, count 16, and so on. At the end of this counting process, you then have a

TABLE 3.1.
*Planetary masses in terms of Earth mass*[a]

| Mercury | 0.054 |
|---------|-------|
| Venus | 0.815 |
| Earth | 1.000 |
| Mars | 0.108 |
| Jupiter | 317.8 |
| Saturn | 95.2 |
| Uranus | 14.5 |
| Neptune | 17.2 |

a. Pluto is not included here, because many astronomers believe that Pluto is not a true planet, but merely a satellite that once escaped from Neptune.

measure of the mass of the body in question, measured in terms of the hydrogen atom as your standard unit. You can imagine this process being carried out for any body, for any planet, or for any star.

This way of thinking about mass is better than using weights, because the weight of a body depends not just on the mass of the body itself, but also on the strength of gravity. The same body would have a different weight on the Moon, because gravity on the Moon is much less than it is on the Earth. So there is nothing absolute about weight determinations, although weights can be used to *compare* masses provided different bodies are weighed under exactly the same circumstances.

Using the counting method for each of the planets, we know how the mass of one planet compares with another. Using the mass of the Earth as the unit of measurement, we have the values given in Table 3.1. The remarkable point emerges that the four inner planets, Mercury, Venus, Earth, and Mars, all have masses much less than the masses of the four outer planets, Jupiter, Saturn, Uranus, and Neptune.

Not only are the masses different for these two planetary groups, but the materials of which the planets are largely constructed are quite different too. The inner group is largely constructed from rocks and from heavy metals like iron, whereas Jupiter and Saturn consist mainly of hydrogen and helium, Uranus and Neptune mainly of water, carbon dioxide, methane, and ammonia. The two groups are so distinct that their separation can hardly be due to accident. These differences must be related to the manner in which the whole system of planets was formed in the first place. Indeed, such differences give astronomers valuable clues when they come to study the problem of how

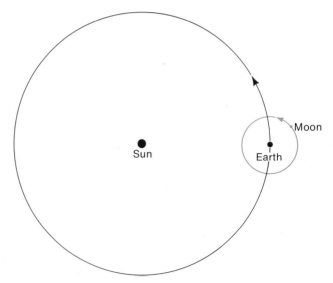

FIGURE 3.1.
The motion of the Moon around the Earth is similar to the motion of the Earth around the Sun. The scale of the Moon's orbit is exaggerated in this figure.

planetary systems are actually born. Nowadays it is generally believed, as a result of such studies, that planetary systems must be widespread, that many of the stars we see in the sky must possess systems of planets quite similar to our own.

The Moon does not appear in Table 3.1, because the Moon is classed as a satellite, not a planet. A satellite is a smaller body moving around a planet, much as the planet itself moves around the Sun. Thus the Earth moves around the Sun, while the Moon moves around the Earth, in the manner of Figure 3.1. The two innermost planets, Mercury and Venus, have no satellites; Mars has two very small ones, much smaller than the Moon; Jupiter has thirteen satellites, four of them quite similar to the Moon; Saturn has nine satellites; Uranus five; and Neptune two. Also, Pluto may be a satellite which has escaped from Neptune.

Although all the satellites have masses less than those of the planets, the mass of the Moon relative to the Earth is 0.012, not greatly different from the value 0.054 for Mercury. Thus the differences in size and mass between the larger satellites and the smallest planet are not substantial; so there is no really firm dividing line between the planets and the satellites. The surface features of Mercury and the Moon are also remarkably similar, as can be seen by comparing Figures 3.2 and 3.3.

The surface features on the Moon have been closely studied in recent years, particularly in connection with the Apollo program. The surface of the Moon is also magnificently accessible to a terrestrial observer equipped with even a small telescope. The most obvious features are the large, more or less circular basins, known as "maria," the general pockmarked, crater-strewn landscape,

FIGURE 3.2.
A recent picture of
the planet Mercury.
(Courtesy of NASA.)

FIGURE 3.3.
In view of the superb photographs now available from space vehicles, we may tend to forget how
very good the old pictures obtained from telescopes on the Earth really were. Compare with Figure 3.2
and note the remarkable similarity. (Courtesy of the Hale Observatories.)

FIGURE 3.4.
An object at left center has rolled downhill on the Moon, leaving a series of marks like prints in a snowfield. (Courtesy of NASA.)

and the absence of apparent movement. Yet some movement is certainly present on the Moon. Figure 3.4 is a picture taken by a space vehicle close to the Moon. A stone, or some object, has plainly rolled downhill, leaving a series of marks rather like steps in a snowslope. And what shall we make of the sinuous meandering channels, to be seen in Figure 3.5? How can these be explained except as caused by the flow of some liquid? Water? No water has unequivocally been detected on the Moon at the present time, but was this always so? Or could there be frozen water locked away below the visible surface? These questions remain unanswered.

By now it is widely agreed that many, if not all, of the craters were formed by objects that struck the Moon from outside. But there are different views concerning the origin of the dark maria. Many astronomers, geologists, and geophysicists believe them to have arisen from episodes of intense volcanic activity in which molten lava, released from the Moon's interior, spread out over the maria basins. Others think the maria basins were also formed by the impact of objects from outside, and that the Moon has always been a mass of cold rock, without volcanoes, without plate movements like those of the Earth. They think the Moon now is essentially the way it was at the time of its formation.

There are several exceedingly peculiar circumstances which bear on this difference of opinion. The Moon always presents very nearly the same aspect to the Earth. There is a part of the Moon's surface, of which Figure 3.6 is an example, which cannot be seen at all by a terrestrial observer. Figure 3.6 belongs to this far side of the Moon. The two sides, "our" side and the far side, are shown adjacent to each other in Figure 3.7. It is

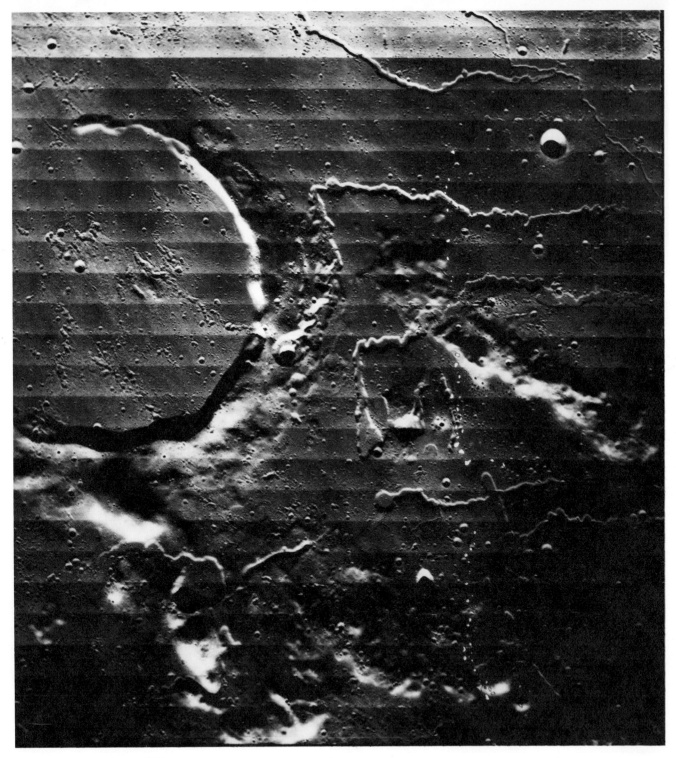

FIGURE 3.5.
These sinuous meandering channels on the Moon suggest a flow at one time of some liquid.
(Courtesy of NASA.)

FIGURE 3.6.
Only from space missions has it been possible to see the far side of the Moon. Note the large double-ringed structure. There are several of these structures on the far side of the Moon, but none on "our" side. (Courtesy of NASA.)

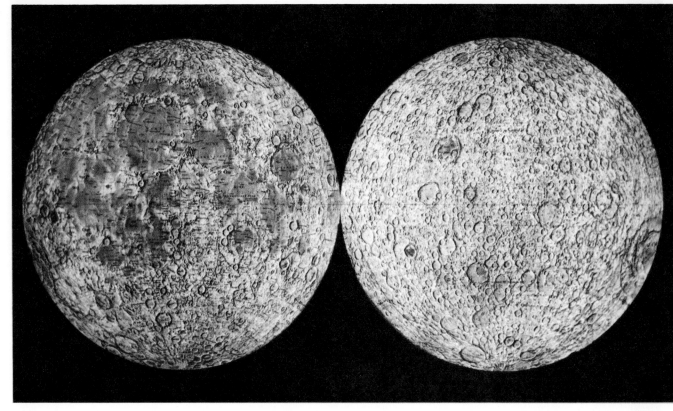

FIGURE 3.7.
The two halves of the Moon. The half we see from the Earth is on the left. There are no dark maria on the far side of the Moon. (Courtesy of Prof. T. Gold, Cornell University.)

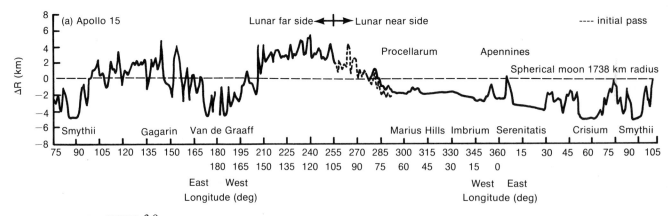

FIGURE 3.8
The far side of the Moon is also much rougher than our side. The jagged peaks and troughs of the altimeter record for the far side are much more marked than on the record for our side. (Courtesy of NASA.)

immediately apparent that there are no large dark maria on the far side of the Moon. How is this to be explained in terms of a volcanic theory? How does molten rock inside the Moon "know" whether it will be facing the Earth when it eventually breaks through to the surface?

Another very queer situation is revealed by Figure 3.8, which is an altimeter record obtained during the flight of Apollo 15. The striking thing about this record is that the difference of elevation between high ground and low ground is about twice as great on the far side of the Moon as it is on the near side. This is only explicable, it seems, if we suppose that the low basins have been submerged on the near side by some kind of "filler" material, by a kind of sea of material out of which only the high ground rises. Although this explanation might seem consistent with lava flows as the filling material, how then do we explain the further fact, also apparent from Figure 3.8, that the high ground is much more spiky on the far side than it is

on the near side? The filler material must have the ability to smooth high ground as well as to fill up the hollows in the low ground. This suggests a much more friable, more powdery material than lava.

Figure 3.9 shows the remarkable case of Mare Orientale. Its origin in terms of an impact theory seems clear. An incoming object, provided it doesn't have too large a speed, produces a pool of molten rocks in the region of impact. This pool is agitated into a circular wave pattern rather like the ripples which continue on a circular pool of water after a stone has been dropped at the center. Eventually, however, the rocks cool to a point where their stiffness stops the wave motion. At this point the wave pattern becomes permanently frozen, and it is the frozen wave pattern we see in Mare Orientale.

Four other, similar wave structures are found on the far side of the Moon, but none on the side facing the Earth. This is the opposite situation from

FIGURE 3.9.
The multiple-ringed structure of Mare Orientale. These are probably frozen waves
from a pool of once-molten rock. (Courtesy of NASA.)

that of the dark maria, suggesting that, if the filler material were removed from the dark maria on "our" side of the Moon, ring structures like Mare Orientale would be found underneath.

A final thought on the Moon. Suppose that, instead of being at the edge of the Moon, Mare Orientale had happened to be exactly at the center of the full moon. What effect would such an "eye" gazing down on the Earth have had on the beliefs of primitive peoples? And how far would those beliefs still affect us today?

Turning back to Table 3.1, we can see that the planet most nearly similar to the Earth in mass is Venus. The diameter of Venus is about 12,200 kilometers, only about 500 kilometers less than the diameter of the Earth. So we expect the inner structure of Venus to be much like the Earth, probably with a similar system of plate movements.

Venus is almost totally cloudbound. Figure 3.10 shows the structure of the clouds. This picture was taken recently from the space vehicle Mariner 10, when at a distance of about 700,000 kilometers from the planet. Television cameras operating in ultraviolet light were used to obtain it.

The clouds of Venus were at one time thought to be like the clouds of the Earth's atmosphere, aggregations of water droplets drawn up from a great ocean by the hot Sun. But we know now that there is exceedingly little water on Venus. For a long time the nature of the clouds remained a mystery. Recently, however, good reasons have been given for believing that they are aggregates, not of water droplets, but of droplets of sulfuric acid.

The temperature at ground level on Venus has been found to be far above the boiling point of water, indeed, to be at about the temperature of molten lead. Evidently the surface of Venus is

FIGURE 3.10.
Venus photographed from a distance of 700,000 kilometers by ultraviolet television cameras (Mariner 10). Winds circulate around the planet in about four days. The clouds may be droplets of sulfuric acid. (Courtesy of NASA.)

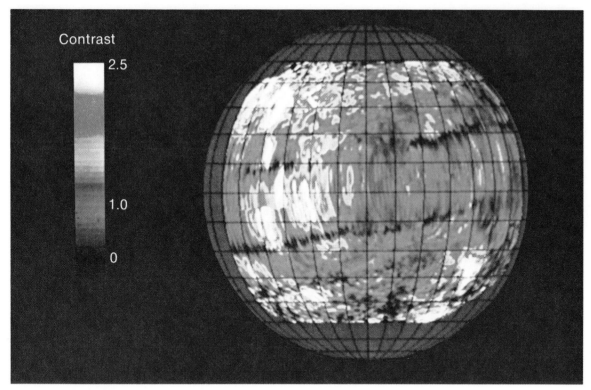

FIGURE 3.11.
Radiowaves penetrate through the clouds of Venus, permitting a map of the surface to be obtained. High ground is indicated by the yellow color and low ground by blue. The topography of Venus appears similar to that of the Earth, except that Venus has large craters like those of the Moon; similar craters on the Earth would have disappeared by now because of erosion. (Courtesy of Dr. D. B. Campbell, Northeast Radio Observatory Corporation.)

totally hostile to life as we know it. Although Venus is similar to the Earth in its gross features of size and mass, the outer atmospheric properties of the two planets are grotesquely different from each other.

Because of the shielding clouds, we cannot see through to the solid surface of Venus in the normal way. Even so, we can form some idea of the surface landscape by bouncing radio waves from the ground. Radio waves transmitted from the Earth travel across space to Venus, where they penetrate the clouds and are reflected back again into space from the solid planet. By the technique known as "radar," we can detect these reflected waves and use them to build up a map of what the solid surface of Venus looks like. The result is shown in Figure 3.11. This figure shows high ground and low ground, and is probably quite similar to what the Earth would be like if the terrestrial oceans were all taken away. There is evidence in Figure 3.11 of large craters like those on the Moon. Presumably such craters, caused by the impact of objects, existed on the Earth at one time, but have since disappeared because of the erosion brought about by water on the Earth.

Mars is often referred to as the "red planet."

FIGURE 3.12.
Ground-based photographs of Mars, showing the effect of a dust storm in the righthand picture. (Courtesy of the Lowell Observatory.)

Figure 3.12 gives a vivid impression of how Mars appears when seen in a telescope under favorable terrestrial conditions. The three panels of this figure, obtained at different times, show that Mars changes markedly in appearance. The picture on the right was taken at a time when an enormous dust storm was raging on the planet.

The surface of Mars has many features similar to those of the Moon, such as large basins and extensive cratering. On the other hand, Mars has turned out to have many features which, before the recent flight of the space vehicle Mariner 9, had not been expected. Figure 3.13 shows a vast and unsuspected canyon, which has an over-all length of more than 2,500 kilometers and an average width of about 150 kilometers. A curious feature of this canyon system is that no place can be seen where material taken from its excavation has been deposited. There are no obvious piles of debris. Perhaps the least radical explanation of this mystery is that

the material which once filled the canyon has been transported away to other parts of the planet by winds. The material would of course need to have been in the form of small grains, and would have been deposited elsewhere in a system of thin layers.

Figure 3.14 is a reconstruction of the whole surface of Mars, shown in three views from three different directions. Wind appears to be the principal agent of erosion on Mars, with sunlight absorbed by dust as the driving agency. The Martian dust storms appear to attain a far greater scale and to be more violent than similar storms on the Earth, a difference coming from the lower density of the Martian atmosphere, which consists largely of carbon dioxide. The lower density leads to higher speeds for the winds. The dust storms are responsible for many of the detailed surface markings, for patterns of dark and light streaks, and for the dune structure seen in Figure 3.15.

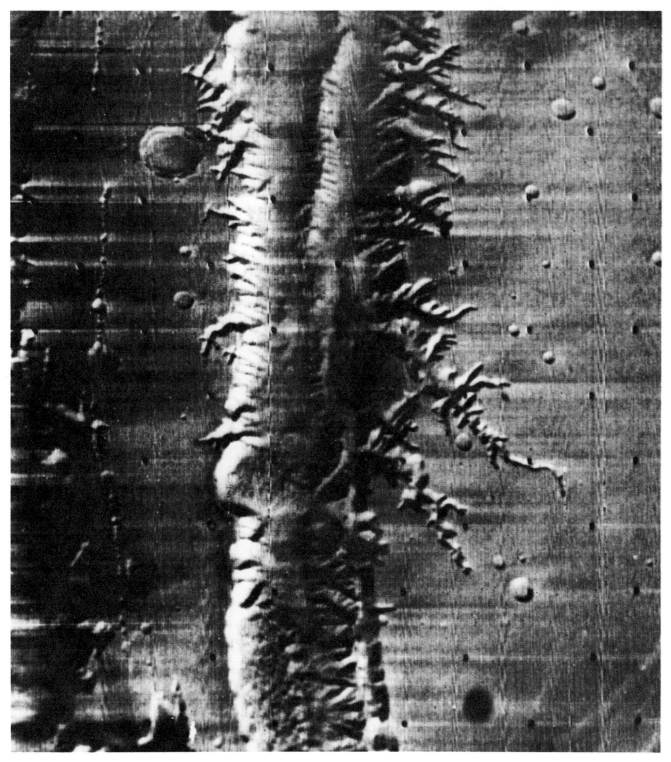

FIGURE 3.13.
A portion of the Coprates canyon of Mars. The canyon has an over-all length of more than 2,500 kilometers and an average width of about 150 kilometers. There appears to be evidence of fluid erosion, showing in the many channels which enter the canyon from the side. (Courtesy of NASA.)

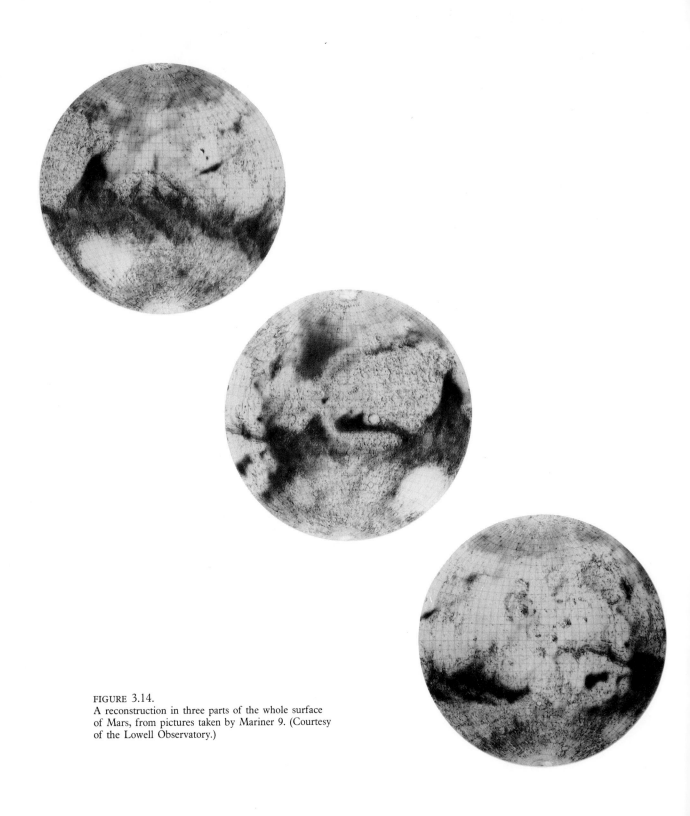

FIGURE 3.14.
A reconstruction in three parts of the whole surface
of Mars, from pictures taken by Mariner 9. (Courtesy
of the Lowell Observatory.)

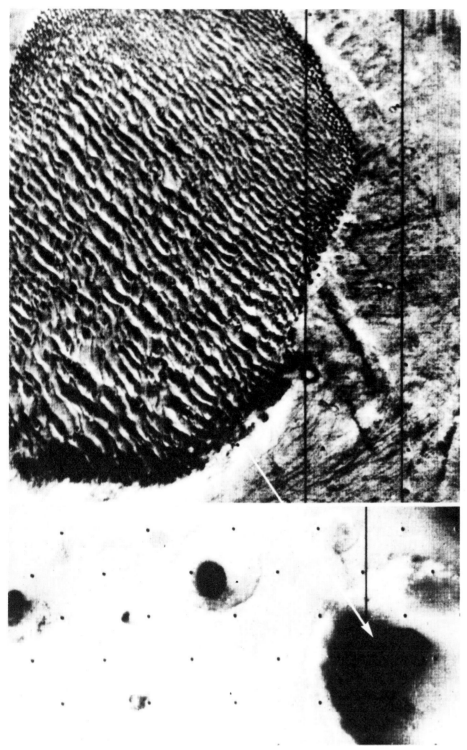

FIGURE 3.15
A dune structure produced by the Martian winds. (Courtesy of NASA.)

The average temperature on Mars is about a hundred degrees Fahrenheit lower than the average temperature on the Earth. At midday on the equator of Mars, ice would reach its melting point, but during the Martian night the temperature falls to more than a hundred and fifty degrees below the freezing point of water. Extensive glaciers of ice could exist on Mars. Because of their permanence, they would become covered by dust deposits carried by winds, and so would not be immediately visible. The situation is different for carbon dioxide. Frozen carbon dioxide (dry ice) has a lower evaporation temperature than ordinary ice, and the dry-ice evaporation temperature lies within the range of Martian variability. In the warmer spots solid carbon dioxide will evaporate into vapor; in the colder spots this vapor will condense into a white frost of dry ice. This circumstance explains the variability of the polar caps, which grow extensively during the Martian winter and which evap-

orate during the summer. It is thought that under a comparatively thin covering of carbon-dioxide frost may lie extensive deposits of ordinary ice—although ordinary ice could persist, and may be present, in the non-polar regions of Mars.

There are features of the surface topography of Mars which seem clearly to have been formed by extensive flows of some liquid, as in the tributaries to the canyon system of Figure 3.13. Such features are suggestive of flash floods rather than the steady flow of a river. The fluid responsible for these features is thought to have been water, and some controversy has occurred over the origin of the water. One possibility is that liquid water may be released from time to time from the interior of Mars, where the temperature is high enough for the water to be in liquid form. Another possibility is that the Martian climate may have long-period cycles, with episodes during which the average temperature is much higher than it is now. In such warmer periods, the glaciers, now frozen, would melt, and liquid water would flow upon the surface of the planet. Whether the desert would bloom is another matter, for it is now thought rather unlikely that there will turn out to be life on Mars.

It is curious that, whereas lunar observers have searched assiduously but without success for signs of recent volcanic activity on the Moon, several very large volcanoes have been found on Mars. Nix Olympica is an enormous, gently rising cone some 600 kilometers in diameter with the central steep-walled caldera shown in Figure 3.16. The height of the center above the general level of the Martian surface has been estimated at 100,000 feet, that is, about 20 miles. The slopes of the rising cone appear to be smooth, being largely free from cratering. This general absence of cratering is interpreted to mean that Nix Olympica is a quite "recent" feature, and may be only a few tens of millions of years old.

It has seemed natural to geologists to interpret these volcanoes in terms of experience on the Earth, although the Martian volcanoes are many times the scale of the largest terrestrial volcano. Thus the outer cone, falling away from the caldera of Figure 3.16, is thought to consist of the usual volcanic material, lava and ash. But this explanation raises the difficulty that such a huge volcano would then need to have astonishingly deep roots. This is because the density of fluid rock is not much less than the density of solid rock, and a great depth of fluid rock would be needed to develop enough pressure to force molten lava to a height of 100,000 feet. This problem would be much less severe if the liquid that formed these volcanoes had had a density much less than that of solid rock. Ordinary water is a possibility for such a liquid. Could these structures really be vast domes of ice, covered smoothly by the deposits laid down by many dust storms? The question is interesting.

FIGURE 3.16.
The central caldera of Nix Olympica. It has been estimated that this central part of the volcano may rise as high as 100,000 feet above the general level of the Martian surface. (Courtesy of NASA.)

Let us turn now from the four inner planets to the four outer ones. Jupiter is shown in Figure 3.17 and 3.18, and the magnificent ring structure of Saturn is shown in Figure 3.19. No comparable pictures exist for Uranus and Neptune. Both Figures 3.17 and 3.18 were obtained by Pioneer space vehicles (10 and 11). A vehicle capable of obtaining similar pictures of Uranus and Neptune had been planned, but after discussions between NASA and the U.S. government, the proposed mission was canceled. It is unlikely that any mission of this kind will actually fly in the foreseeable future. Because of their great distance, Uranus and Neptune appear only as small objects when seen from the Earth, even when a large telescope is used, as can be seen from Figures 3.20 and 3.21.

We have already noted that Jupiter, Saturn, Uranus, and Neptune are much larger than any of the four inner planets. Yet in spite of their greater sizes, the outer planets rotate rapidly on their polar axes, in periods which range from about 9 hours 50 minutes for Jupiter to 15 hours 40 minutes for Neptune. None of the inner planets spin as fast as this. It is indeed these fast rotations which cause Jupiter and Saturn in Figures 3.17, 3.18, and 3.19 to be appreciably flattened from a spherical shape.

The gas methane, known colloquially as "marsh gas" or sometimes as "fire damp," is found in the atmosphere of all four outer planets. Methane is the simplest of the hydrocarbons—oil being formed when molecules of methane are joined together in a suitable way. It is ironic, since we attach such great financial and political importance to the comparatively small quantities of oil found on Earth, that there should be incomparably vaster quantities of "natural gas" in the atmospheres of all the outer planets. Ammonia in gaseous form is

FIGURE 3.17.
Jupiter with a satellite in transit. This photograph was taken by Pioneer 10 at a distance of about 2.5 million kilometers from the planet. The large oval marking, seen near the fainter part of the limb, is known as the Red Spot. (Courtesy of NASA.)

FIGURE 3.18.
Another view of Jupiter, one that cannot be seen at all from the Earth, showing convection cells near
Jupiter's north pole. These cells rise like thunderstorms do on Earth. (Courtesy of NASA, Pioneer 11.)

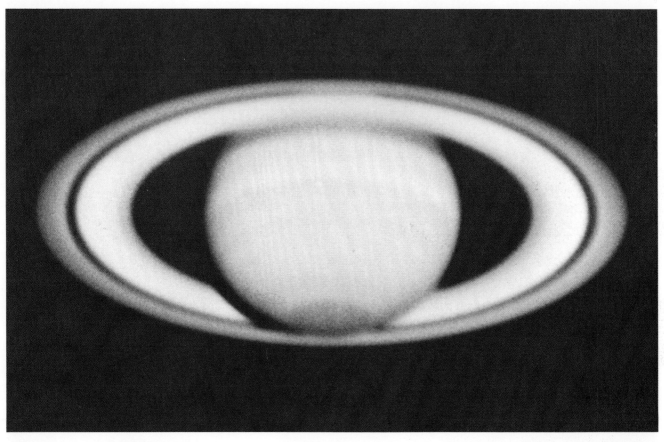

FIGURE 3.19.
The planet Saturn as seen in early 1973. The prominent dark gap in the ring is known as "Cassini's division". (Courtesy of New Mexico State University.)

FIGURE 3.20.
The planet Uranus and three of its satellites.
(Courtesy of the Lick Observatory.)

FIGURE 3.21.
The planet Neptune and the satellite Triton.
(Courtesy of the Lick Observatory.)

found in considerable quantity in Jupiter, in smaller quantity in Saturn, but in Uranus and Neptune ammonia is probably frozen into solid crystals. Hydrogen and helium are present in large quantities in the atmospheres of all four outer planets.

What are the rings of Saturn made of? Probably from spheres of ice several feet in diameter, embedded in a swarm of fine snowflakes. The spheres of ice are particularly effective in reflecting radio waves from the Earth, when the radar technique is used on them.

In Chapter 2 we considered the nuclear engine within the Earth, an engine responsible for plate movements, for the building of mountain ranges, for the occurrence of earthquakes and of volcanoes. This engine produces heat, which gradually works its way to the Earth's surface and which then escapes into outer space. The energy of this heat has been measured, and has been found to be very much less than the energy of the light which the Earth receives from the Sun. Jupiter also contains an engine, probably of a gravitational rather than a nuclear kind. The engine in Jupiter is comparatively much more powerful than that of the Earth, for it supplies *more* energy than all the sunlight falling on Jupiter. It seems, then, that the interior of Jupiter is likely to be subject to even more drastic effects than the Earth, and that many of the surface markings on Jupiter, to be seen in Figures 3.17 and 3.18, may be due to huge disturbances within the planet itself.

# 4

## *The Sun*

# 4

## *The Sun*

We found that the largest planet, Jupiter, has a very much greater mass than the smallest planet. In turn, the Sun has a very much greater mass than Jupiter, about a thousand times greater. The relative sizes of the planets compared to the Sun are shown in Figure 4.1. Clearly the Sun is the overwhelmingly dominant body in our system, both in mass and in size.

The Sun is a nuclear furnace, producing its energy by converting atoms of hydrogen into atoms of helium, four atoms of hydrogen going to make one atom of helium. The processes whereby this occurs have been studied by physicists. They can be reproduced on the Earth, but not on anything like a commercial scale. In order to be able to produce energy in useful amounts in this way, we would need to raise hydrogen gas to a very high temperature and *to maintain it there*. The trouble on the Earth is that, whenever we make a gas very hot indeed, it simply blows apart in an explosion.

The hot gases inside the Sun do not blow apart because they are held together by gravity. The Sun contains so much material that gravity within it is strong. On the Earth it is not possible to reproduce this condition, because the gravity created by even the mass of the whole Earth is not strong enough to restrain gases that are as hot as they are inside the Sun. Instead of gravity, scientists have tried to use magnetic forces to solve this problem, but this has turned out to be difficult. (Governments throughout the world sponsor efforts along these lines, for the obvious reason that the energy yield from the conversion of hydrogen to helium would be enormous, if only a way could be found to do it successfully. The hydrogen in a mere thousand tons of ordinary water would, by being converted to helium, provide enough energy to meet the needs of the whole of the world's industry for a year.)

So what the Sun does very easily, man finds very

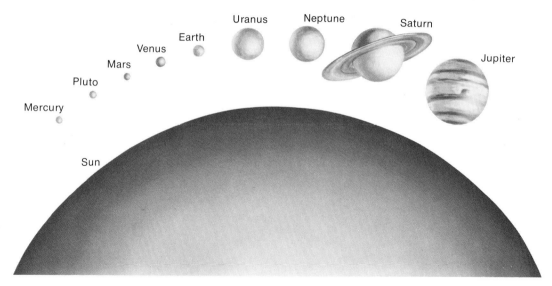

FIGURE 4.1.
The nine planets of the solar system in scale relative to the Sun. (After Gilluly, Waters, and Woodford, *Principles of Geology*, 3d ed. W. H. Freeman and Company. Copyright © 1968.)

difficult. The problem may be solved in the future, but there is no certainty that it will be. A much simpler way of obtaining energy is to use natural sunlight as it falls on the Earth. The sunlight falling on the roof of a house gives sufficient energy for the needs of the household, but there is the difficulty that the heat of summer must be stored for use in winter. Modern science offers nothing better than storage in expensive and cumbersome electric batteries, which are much too inefficient to be considered a satisfactory solution. The best way to make use of summer heat in winter is still the age-old method of burning in winter the wood that was grown in the summer.

A less direct way of using sunlight is by means of the water that evaporates from the oceans. The water vapor rises from the ocean surface to considerable heights in the atmosphere. Some of it falls as rain and snow on high ground, where it can be trapped in lakes. The water from such high lakes can then be run downhill and made to drive electrical machinery. This hydroelectric power would be adequate to run the world's industry if the human population were, say, about a quarter of its present size.

When viewed in full sunlight, the Sun appears as in Figure 4.2. With even a small telescope, one can often see dark, more or less circular, spots. These *sunspots* were discovered by Galileo (1564–1642; Figure 4.3), whose drawings of them are shown in Figure 4.4. Galileo's small telescope is shown in Figure 4.5. It is interesting to compare Figure 4.5 with the modern, 200-inch Hale telescope shown in Figure 4.6. This difference brings out the technological gulf between our present-day world and the society of four centuries ago, when the world's population was not far from 500 million.

The discovery of sunspots caused a great sensation, since the face of the Sun was previously

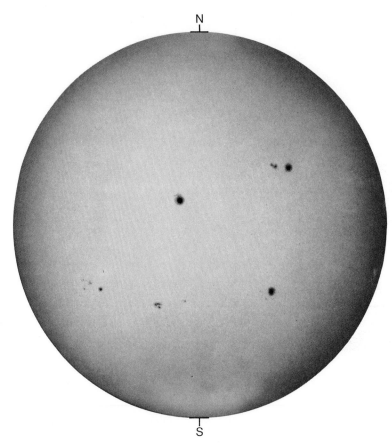

FIGURE 4.2.
Viewed in full sunlight, spots can be seen on the Sun, contradicting the ancient belief that the fact of the Sun must be perfect.
(Courtesy of Dr. N. Sheeley, Kitt Peak National Observatory.)

FIGURE 4.3.
Galileo Galilei (1564–1642).
(Courtesy of the Roman Picture Library, Cambridge.)

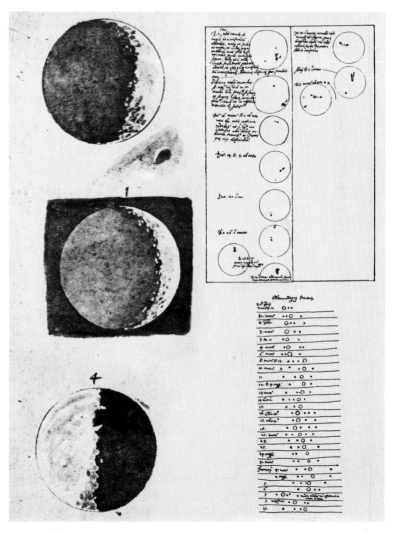

FIGURE 4.4.
Galileo's observations of sunspots. Drawings from a notebook.

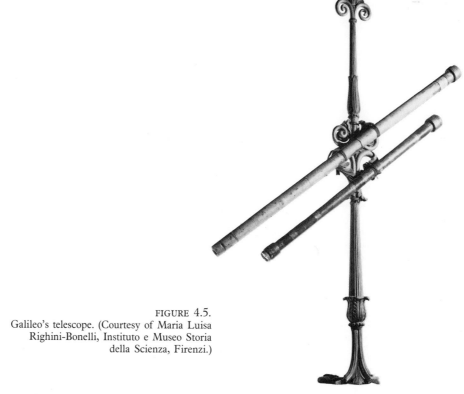

FIGURE 4.5.
Galileo's telescope. (Courtesy of Maria Luisa
Righini-Bonelli, Instituto e Museo Storia
della Scienza, Firenzi.)

FIGURE 4.6.
The 200-inch Hale Telescope.
(Courtesy of Hale Observatories.)

thought to be "perfect," a belief which had probably survived from stone-age times, from an era of sun worship. Under favorable conditions large sunspots are easily visible to the naked eye; so it is surprising that their discovery had to await the dawn of modern science. On two occasions I have myself been able to observe such naked-eye spots, even in such an astronomically unfavorable climate as that of the United Kingdom. Naked-eye spots are best seen near dawn or sunset, when absorption in the atmosphere cuts the normal glare of the Sun down to a level at which one can look directly at the solar disk. Usually turbulence in the atmosphere destroys clarity for low angles above the horizon, leaving the setting Sun as a "boiling" image. In order for large sunspots to be seen by the naked eye, the atmosphere must give absorption but without distortion, and this condition is encountered only rarely. Nevertheless, naked-eye spots must have been seen long before Galileo,

probably by many thousands of people. The surprise is that their existence does not seem to have been recorded by monks or by court chroniclers. Sunspots are dark only in contrast with the brilliance of the surrounding regions of the solar surfaces—the temperatures within sunspots are actually higher than that within most industrial furnaces.

The appearance of the Sun undergoes a dramatic change when, instead of being photographed in ordinary white light, it is photographed by means of light from particular kinds of atoms. Figure 4.7 is a photograph of the Sun in the light characteristic of hydrogen atoms, and Figure 4.8 is in light characteristic of calcium atoms. The remarkable structures revealed in these pictures suggest the presence of unusual forces. These forces are thought to be magnetic, the same kind of forces which terrestrial scientists are using in their quest for nuclear power. The effect of magnetic forces

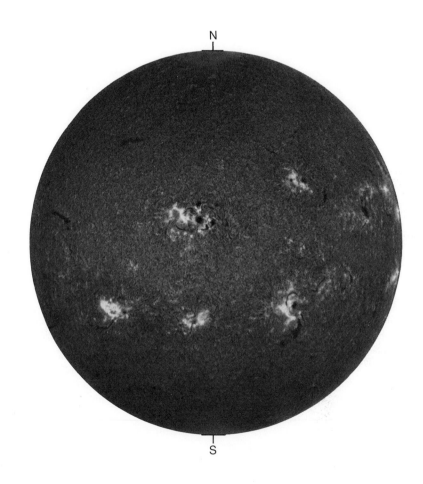

FIGURE 4.7.
The Sun, as in Figure 4.2, but photographed in the light emitted by hydrogen atoms (Hα). (Courtesy of Dr. N. Sheeley, Kitt Peak National Observatory.)

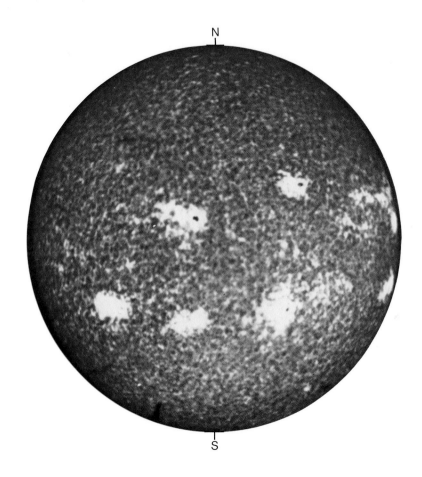

FIGURE 4.8.
The Sun, as in Figure 4.2, but photographed in the light emitted by calcium atoms (K line).
(Courtesy of Dr. N. Sheeley, Kitt Peak National Observatory.)

FIGURE 4.9.
The solar corona at the total eclipse of June 1973. The form of the corona is indicative of the presence of magnetic forces. (Courtesy of the High Altitude Observatory.)

can also be seen in the white-light photograph of Figure 4.9, which was obtained during an eclipse of the Sun. Eclipses occur when the Moon comes directly between the Earth and the Sun, as in Figure 4.10, thereby cutting off the light from the solar disk, and permitting the tenuous gas which exists around the Sun to be seen. The gas in this extended region, or "corona" as it is called, has turned out to be surprisingly hot. However, because it is so tenuous, no great amount of energy is emitted by it. Nevertheless the emission by this gas is extremely interesting, for much of it occurs not as ordinary light, but as x-rays. A picture taken with these x-rays is shown in Figure 4.11. Notice the presence of remarkable "hotspots," related

to the positions of sunspots.

Sunspots wax and wane in their numbers in a cycle of about 11 years. Other phenomena also vary with the sunspots. Changes in the shape of the corona, particle streams ejected out through the corona, giant prominences in the solar atmosphere of the kind shown in Figure 4.12, all vary cyclically with the sunspots. Particle streams are associated with "flares," which appear to be localized regions in the lower solar atmosphere where intense electrical discharges take place. The discharges are accompanied by heating, causing flares to appear bright against the solar disk, as in Figure 4.13. The origin of sunspots, of flares, of prominences, of the corona, and of the solar cycle, is

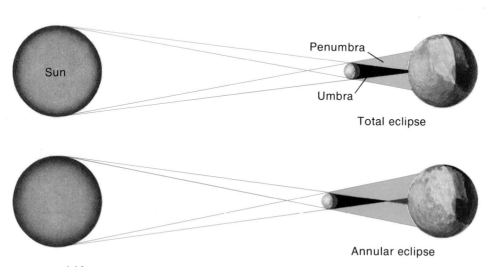

Penumbra

Umbra

Total eclipse

Annular eclipse

FIGURE 4.10.
At a total eclipse of the Sun, the Moon's direct shadow reaches the Earth. (From J. Brandt and S. Maran, *New Horizons in Astronomy*. W. H. Freeman and Company. Copyright © 1972.)

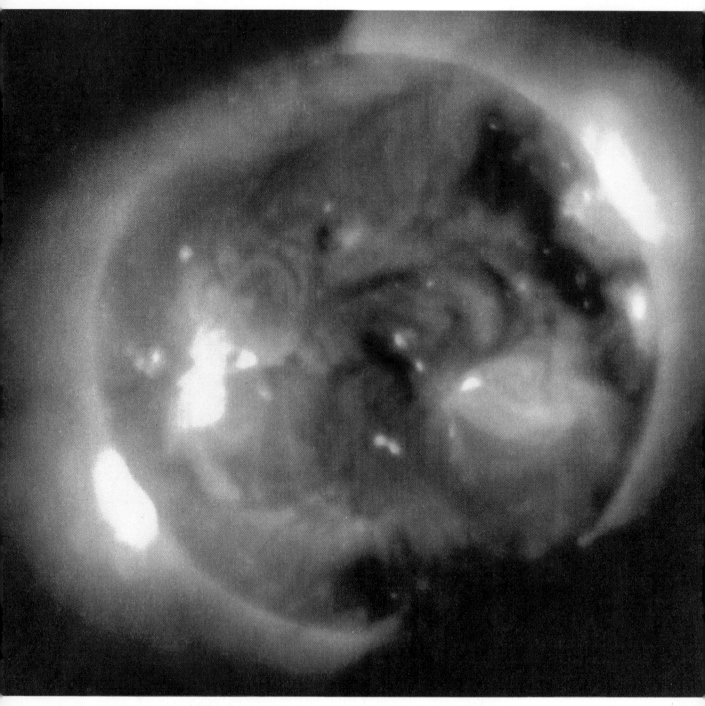

FIGURE 4.11.
X-rays emitted by the hot corona of the Sun.
(Courtesy of American Science and Engineering, Inc.)

FIGURE 4.12.
A giant prominence on the Sun. Gases in the corona are cooling and moving under the influence both of gravity and of magnetic forces. (Courtesy of the Hale Observatories.)

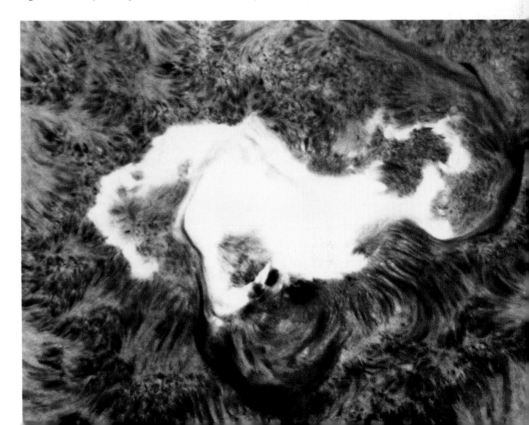

FIGURE 4.13.
A flare on the Sun. Flares cause streams of particles to be ejected from the Sun. Such jets of particles move rapidly outward and sometimes impinge on the Earth. (Courtesy of the Hale Observatories.)

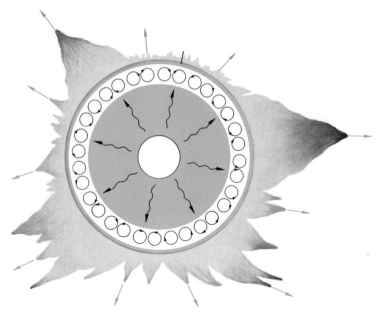

FIGURE 4.14.
A schematic representation of the structure of the Sun. Energy is generated in a central core. The energy is then carried by radiation to a subsurface zone, where it produces convective motions which may well be responsible for the solar cycle. (From J. Brandt and S. Maran, *New Horizons in Astronomy*. W. H. Freeman and Company. Copyright © 1972.)

probably associated with the fact that the Sun, immediately below its visible surface, is in a violently boiling condition.

The internal structure of the Sun is shown in Figure 4.14. The nuclear-energy production we discussed above occurs in a hot inner core. This energy generation causes the immediate subsurface regions to take up a circulating convective motion, rather like the contents of a saucepan heated from below. The cycle of the sunspots and of other solar pheomena is probably to be explained in terms of an interplay of magnetic forces with this convective motion. The problems involved seem to be among the most difficult in astronomy.

Let us return our attention to the extensive outer atmosphere of the Sun. The gas of the corona is hot enough that it has no clearly defined outer boundary. A "wind" of particles is constantly streaming outward from the corona. Indeed, such streamers are plainly seen in Figure 4.9. The wind is exceedingly tenuous by ordinary standards—a cubic inch would contain only a few hundreds of atoms. Nevertheless, the wind at its stronger moments can have effects on the Earth, the Moon, and on the other planets. These winds are capable of disturbing the magnetic field of the Earth, for example producing what are known as "magnetic storms."

Why are the diffuse gases of the corona so hot? The answer to this question appears to be rooted in the convective motions in the subsurface regions of the Sun, the motions of Figure 4.14. These motions generate waves that travel outward into the solar atmosphere, getting more violent as the atmospheric gases become more tenuous. In other words, the particles within the gas are constantly being shaken backwards and forwards by these waves, the shaking motion becoming more and more intense as the density of the particles decreases—which it does as the waves rise up into the solar atmosphere. Collisions of one particle with another eventually cause the wave motions to be dissipated into heat, the heating effect being strong where the motions are most violent, in the corona. The process is very likely a good deal more complicated than it may seem from this simple description. Magnetic forces are probably also involved, and many problems remain for precise solution.

I would like to end this chapter on a personal note. I have always found it difficult to believe that the great variety of unusual events in the atmosphere of the Sun could arise from the comparatively slow-moving convection of material below the solar surface. From what at first sight seems a very ordinary situation, a vast web of intricate processes comes to be fashioned. My failure of understanding arose, I now realize, from a lack of attention to a remarkable circumstance which one learns in the study of the somewhat mysterious science of thermodynamics. Consider a gas, say, at a temperature of ten thousand degrees. The average speed of motion of a hydrogen atom at this temperature is about 15 kilometers per second. Some hydrogen atoms have speeds above this average, but only one atom in a mass comparable to that of the whole Sun would have a speed as high as 150 kilometers per second, and none would have a speed approaching that of light, 300,000 kilometers per second. Given only a hot gas at a temperature of ten thousand degrees, there is no way to obtain particles moving at speeds approaching that of light, or indeed to obtain phenomena of the kind found in the solar atmosphere—the x-ray emission of Figure 4.11, for example. But let the gas vary slightly in temperature, say, with one part of it at ten thousand degrees and another part at ten thousand one hundred degrees, and the situation now becomes quite different. In such a situation, there can be a "machine" generating mechanical motion, like the turning of a wheel or the lifting of a weight. Once a heavy wheel can be turned, electricity can be generated. With electricity available, all manner of complex devices become possible. The symphonies of Beethoven can be played on a hi-fi system. Physicists can now build devices which produce particles moving at speeds close to that of light. A maze of intricate and remarkable phenomena become possible, all from the slight difference of temperature in the two parts of our gas.

I have known these facts from my student days. I have even taught classes in thermodynamics, and yet I have always found it difficult in my thinking to take account of the rich and varied occurrences that may arise when two large masses of gas interact with each other. This is just the situation that occurs when convective motions develop within the body of the Sun.

# 5

*Comets and Other
Forms of Debris*

# 5

## *Comets and Other Forms of Debris*

It is generally believed that the four inner planets, Mercury, Venus, Earth, and Mars, were formed from a very large number of much smaller solid bodies, which collided from time to time among themselves, some growing bigger and some being broken up by the collisions. The process went on, probably for several million years, until a few large hunks of material began to dominate over all the others. It was these few hunks which eventually emerged as the inner planets. The cratering to be seen on Mercury, Mars, the Moon, and Venus occurred in the last phases of this aggregation process. The cratering was caused as the planets continued to sweep the solar system clear of smaller pieces of debris.

Similar aggregation probably also occurred for the outer planets, Jupiter, Saturn, Uranus, and Neptune, although growth by the addition of gas, particularly hydrogen and helium, must also have been important for Jupiter and Saturn. The process for the outer planets must have taken much longer to complete than for the inner planets, perhaps as long as 500 million years—about a tenth of the age of the solar system.

We can estimate the sizes of the chunks of material that were added to the planets, during the late stages of this process of debris-sweeping, from the scale of the craters which the debris produced, for example, when the Moon was hit. The size of the object that produced a crater of given dimensions is calculated as a kind of average between the depth of the crater and its diameter. The largest basins on the Moon were very likely produced by chunks of material with a diameter of about 100 kilometers. Objects of this size must have been rare, however. The multitide of small craters on the Moon were produced by the impact of very much smaller bodies, many of them no more than a few meters in size.

These considerations prompt the question: Is

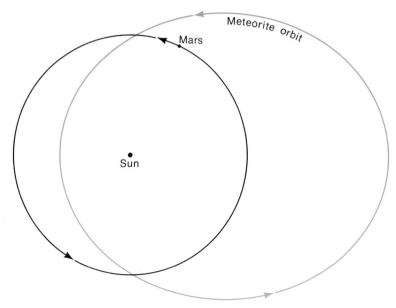

FIGURE 5.1.
A meteoritic fragment comes to follow an orbit which dips inside that of Mars, so that a close approach to Mars becomes possible.

the solar system now swept entirely clean, or does some of the original debris still survive? The answer is that some debris does still survive, in the form of a swarm of bodies moving around the Sun, mostly in the general region between the orbits of Mars and Jupiter. These bodies, known as asteroids, are all very much smaller than the Moon. The largest of them, the asteroid Ceres, has a mass that is about 1 per cent of the mass of the Moon. The largest ten or so of the asteroids have an average diameter of about 300 kilometers, so that they are bodies of the general order of size of those which seem to have blasted the largest basins on the Moon.

Sometimes, perhaps as a result of a collision of one asteroid with another, a lump of material may be thrown into a markedly elliptical orbit, one that may even dip inside the orbit of Mars, as indicated in Figure 5.1. The possibility now arises that such a lump may eventually hit Mars and so

become "swept out." Another possibility, a more likely one, is that the lump of material will come close to Mars, but without experiencing an actual collision. The gravitational pull of Mars will then change the orbit once again, perhaps causing it to become still more elliptical, as in Figure 5.2. The possibility now arises, since in Figure 5.2 the orbit of the material dips inside the orbit of the Earth, that it may collide with the Earth.

The process just described is actually going on all the time. It almost always involves comparatively small pieces of material, a few inches or a few feet in diameter, not the rare, comparatively large pieces. Such smaller pieces of material do indeed collide with the Earth. As they travel at high speed through our atmosphere, they produce a blaze of light. Some of them eventually land on the Earth's surface, whence they may be recovered and examined. They are known as *meteorites*, and their study forms an important link between the

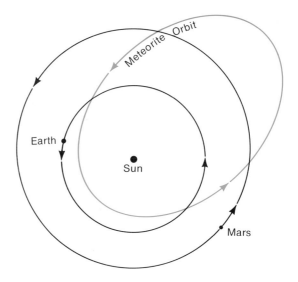

FIGURE 5.2.
The orbit of a meteor can be changed further by an approach to Mars, so that it dips inside that of the Earth. Direct collision with the Earth now becomes possible.

sciences of astronomy and chemistry, a study sometimes referred to as astrochemistry. From them much can be learned about the nature of the material out of which the inner planets were formed some 4,500 million years ago.

Very rarely, perhaps once in 10,000 years, a much larger piece of material strikes the Earth, a piece about a kilometer in diameter. On such rare occasions, a crater of appreciable size is blown in the Earth's surface. If the crater is formed in a dry desert region of the Earth, as near Winslow, Arizona (Figure 5.3), the crater may survive for a considerable period. Otherwise it will quickly be eroded away.

The asteroids are very likely debris left over from the accumulation process which led to the origin of the four inner planets, Mercury, Venus, Earth, and Mars. Debris must also have been left over from the formation of the outer planets. It is

probable that the *comets* are pieces of this other debris.

Perhaps no astronomical object captures popular attention more strongly than the comets. After seeing two brilliant ones myself, both in the year 1957, Comets Arend-Roland and Mŕkos, shown in Figures 5.4 and 5.5, I can understand why they do so. At its most brilliant, such a comet covers a substantial area of the sky, from an arc on the horizon to the zenith directly above. Such a comet appears rather like a luminous vertical sword with the point raised high above one's head. Probably because of this, comets were regarded in the Middle Ages as the harbingers of a disaster, perhaps even of the end of the world itself.

Cometary orbits are mostly of highly elongated form. A typical comet begins its journey a great distance away from the Sun, far out beyond the planet Pluto. It may take thousands, or even hun-

FIGURE 5.3.
The crater near Winslow, Arizona, formed by the impact of a substantial meteorite.
(Courtesy of the American Meteorite Museum)

dreds of thousands, of years to fall toward the inner part of the solar system. A comet typically has a central core of icy material about 10 kilometers in diameter. When the orbit of a comet dips inside the orbits of the inner planets, a collision may occur. Calculation shows that this would be likely to happen to the Earth once in every few million years. The collision would be violent, and would cause extensive devastation. Such collisions could also lead to a quantity of water being temporarily added to an otherwise dry body like the Moon. It is an interesting question whether the sinuous channels (Figure 3.5) to be seen on the Moon's surface could have been caused by water that was added in this way.

It may be wondered how an icy chunk of material only some 10 kilometers in diameter, very small compared to the scale of the Earth's orbit around the Sun—the average distance of the Earth from the Sun is 149.6 *million* kilometers—can show itself as an object spread across a substantial fraction of the sky. This question is answered by the volatility of much of the initially solid material of the comet. As a comet comes toward the Sun, a fraction of its material evaporates into a gaseous form, a process which usually begins when the comet moves in

FIGURE 5.4.
The comet Arend-Roland, April 29, 1957. (Courtesy of the Hale Observatories.)

FIGURE 5.5.
The comet Mrkos, August 22, 1957. (Courtesy of the Hale Observatories.)

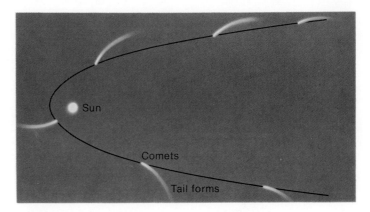

FIGURE 5.6.
A "tail" forms as a comet approaches the Sun. Throughout the passage of the comet, the tail always points *away* from the Sun.

toward the orbit of Mars. The process becomes stronger still as the Sun is approached more closely, particularly for the occasional comet that goes inside orbit of Mercury.

The gases, as they evaporate away from the solid nucleus, form what is known as the "head" of the comet. The head is often affected by light from the Sun and also by the wind of particles from the Sun. The processes are complex, but they cause a portion of the gas to leave the head altogether. This gas moves into a "tail," which tends to point outward, away from the Sun, as in Figure 5.6.

The gases moving outward in a cometary tail are not recovered. Indeed, it is just this streaming away of the tail which produces the vast size of a comet. Usually the length of the visible tail—visible because sunlight is re-radiated by the gas—is about one-tenth of the scale of the Earth's orbit around the Sun. However, for spectacular comets the visible tail can be as large as the Earth's orbit itself. It

is the development of the tail which gives so impressive a scale to spectacular comets.

If the solid material of a comet were influenced only by the gravitational pull of the Sun, the remains of the icy nucleus would follow a highly elliptical orbit, taking the nucleus back to the extreme outer regions of the solar system from where it came. It is a property of such a highly flattened elliptical orbit, however, that the gravitational pull of a large planet like Jupiter can change it quite markedly. Thus, mainly because of Jupiter, comets do not return to their starting positions.

The orbit of a comet can be altered decisively in two quite different ways, depending on where Jupiter is in relation to the comet's orbit. The orbit may be changed so that the comet leaves the solar system altogether, and moves out into the spaces between the stars. The other possibility is illustrated in Figure 5.7. Here the comet does not retreat to the great distance from which it came,

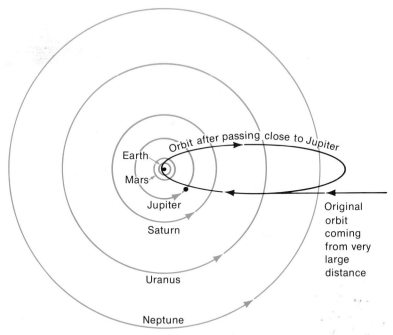

FIGURE 5.7.
A comet in an orbit with a very long period becomes perturbed by the gravitational influence of Jupiter into an orbit of much shorter period. (Adapted from J. Brandt and S. Maran, *New Horizons in Astronomy*. W. H. Freeman and Company. Copyright © 1972.)

although it still retreats to the region of the outer planets. Instead of taking hundreds of thousands of years to go around the Sun, the comet now completes a circuit of the Sun in only a few scores of years. The period for Halley's comet shown in Figure 5.8, is 76.2 years.

Newton's theory permitted highly elliptical orbits as well as nearly circular ones. None of the planets had highly elliptical orbits, and this lack was explained in terms of the way the planets were set moving in the first place. But could other, non-planetary bodies be found which did follow highly elliptical paths around the Sun? In Newton's time this was an important question. Comets were possibilities. Edmund Halley in 1705 published orbits of more than 20 comets which had been observed during previous centuries and for which he thought the records reliable. He found three of them to be similar to each other, those for the comets observed in the years 1531, 1607, and 1682. The even spacing of about 76 years between these apparitions suggested that the same body might have been sighted all three times. If this was so, in a further 76 years the comet would return again.

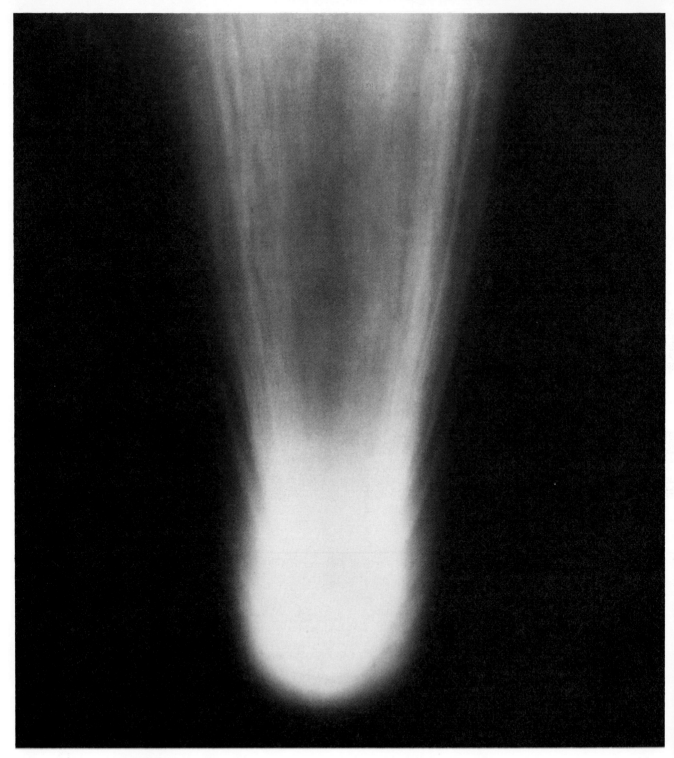

FIGURE 5.8.
Halley's comet was probably perturbed into its present orbit, with a period of 76.2 years, by the influence of Jupiter. (Courtesy of Hale Observatories.)

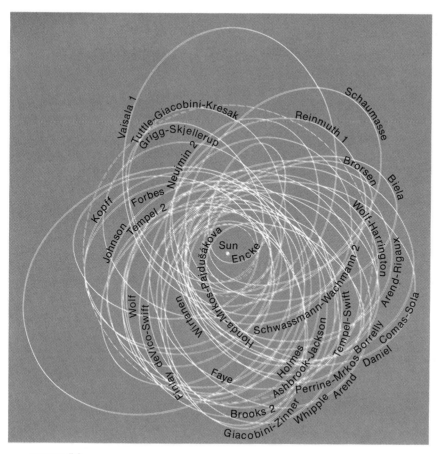

FIGURE 5.9.
The orbits of short-period comets lie mostly in the region between Mars and Jupiter. This is the graveyard of the comets.

When the year 1758 eventually arrived, "Halley's comet," as it had become called, was eagerly sought. It was, in fact, observed on Christmas night of that year. The last return of Halley's comet to the region of the Earth was in 1910. The next will be in 1986.

Once a comet has become disturbed as much as Halley's comet has, it must experience further disturbances, particularly from Jupiter. The gradual effect of these disturbances will be to round the orbit more and more. Halley's comet will eventually move in an orbit with its greatest distance from the Sun much less than at present; it will no longer move out to the region between Jupiter and Saturn. The eventual orbit will have a scale more comparable to its distance of nearest approach to the Sun, the scale of the orbit of Mars.

The graveyard of the comets lies in the general region between Mars and Jupiter. At present there are about 20 well-known comets in this region,

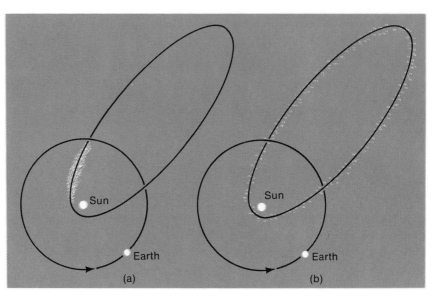

FIGURE 5.10.
A comet breaks up by particles gradually spreading around the whole orbit, as in the change from (a) to (b).

moving around the Sun in periods of about five years. They have the complex system of orbits shown in Figure 5.9. When this stage is reached, most or all of the original icy material has been evaporated, and the resulting gas has been lost. These short-period comets, as they are called, have therefore lost the ability to grow spectacular tails. A comet, as it approaches the last phase of its life, consists of a swarm of refractory particles, which probably range in size from tiny sub-pin-head grains up to substantial chunks of material. Originally all these refractory particles may have been frozen into the volatile ices, like the dirty ice at the snout of a terrestrial glacier. Such comets then suffer the further indignity that their solid particles, particularly the tiny ones, are pulled more

and more away from any nucleus that may remain, so that the small particles come to be spread into a kind of tube, as in Figure 5.10, around the whole of the orbit.

Sometimes the orbit of a comet comes moderately close to that of the Earth. Although the chance of a collision with the Earth is small so long as the comet remains a compact object, collisions become more and more probable as the comet spreads into the tube of Figure 5.10. In the situation of Figure 5.10, a swarm of small particles is strewn around the whole orbit. Such particles strike the terrestrial atmosphere whenever the Earth comes close to the path of the comet. Being tiny and moving at high speeds, these particles are almost immediately evaporated on entering our

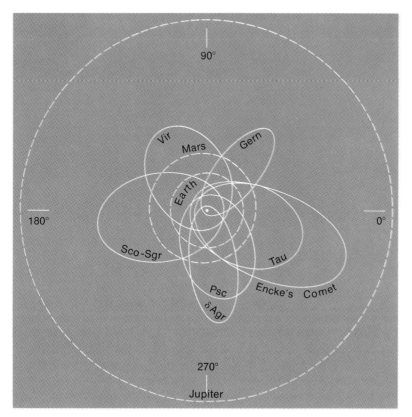

FIGURE 5.11.
Meteor streams in orbits close to the plane of the Earth's orbit around the Sun. These streams have resulted from the gradual break up of comets that once moved in similar orbits, as Encke's comet is here shown to do.

atmosphere. The heat generated by the motion produces a trail of hot gas often visible at night from ground level. Such momentary streaks of light across the sky are called meteors (different from meteorites) or sometimes "shooting stars."

The Earth crosses about a dozen cometary streams, as illustrated in Figure 5.11. It does so at various parts of its orbit, that is, various times in the year. The best months for observing meteors are August and November.

The comets are believed to be debris left over from the primordial process in which the Sun and planets were formed. When we see the momentary flash of a meteor in the sky, we might remember that the particle responsible for it was an old particle, coming to us down the ages from the episode in which the Earth itself was born.

# 6

*Stars*

Just as we can show places on the surface of the
Earth in the form of maps, so maps of the stars in
the sky can be made, as in Figures 6.1 to 6.6.
Different bits of the sky are described by names,
just as countries on the Earth have different names.
The various regions of the sky are called constellations, a complete list of which is given in Table 6.1.

# 6

## *Stars*

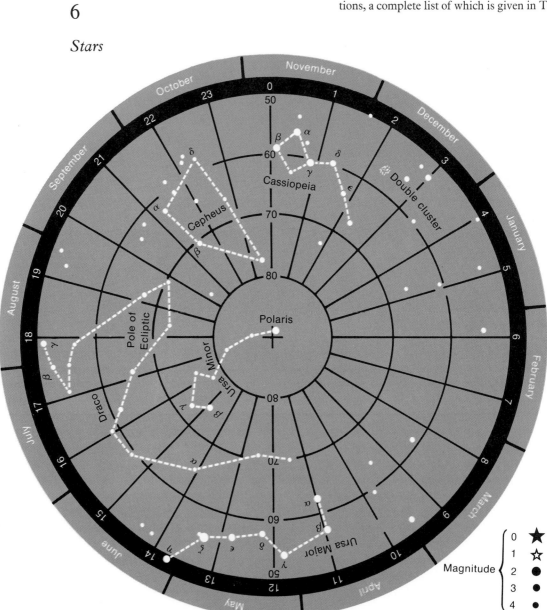

FIGURE 6.1
The constellations of the northern polar cap.

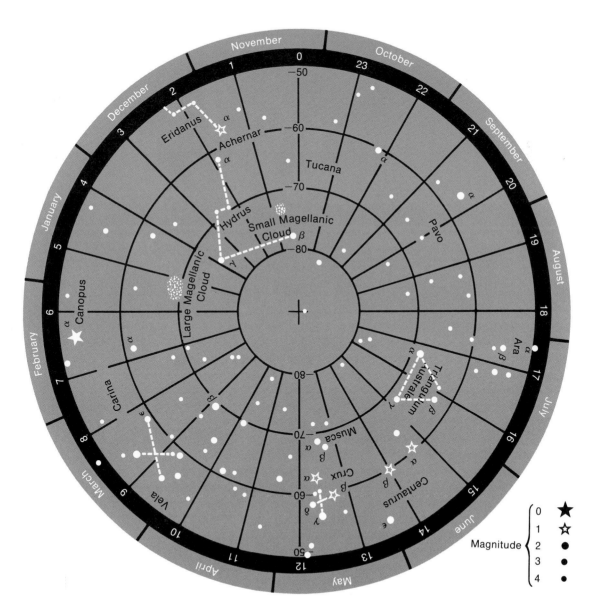

FIGURE 6.2.
The constellations of the southern polar cap.

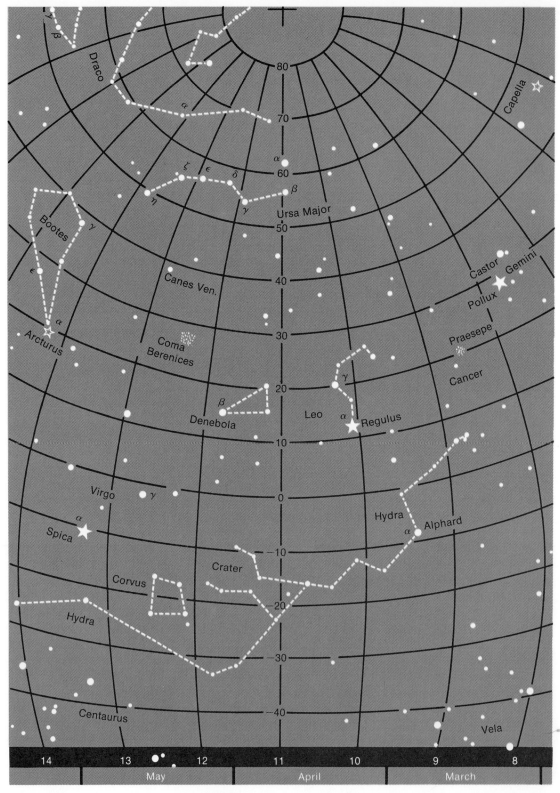

FIGURE 6.3.
Night sky in Spring (northern hemisphere).

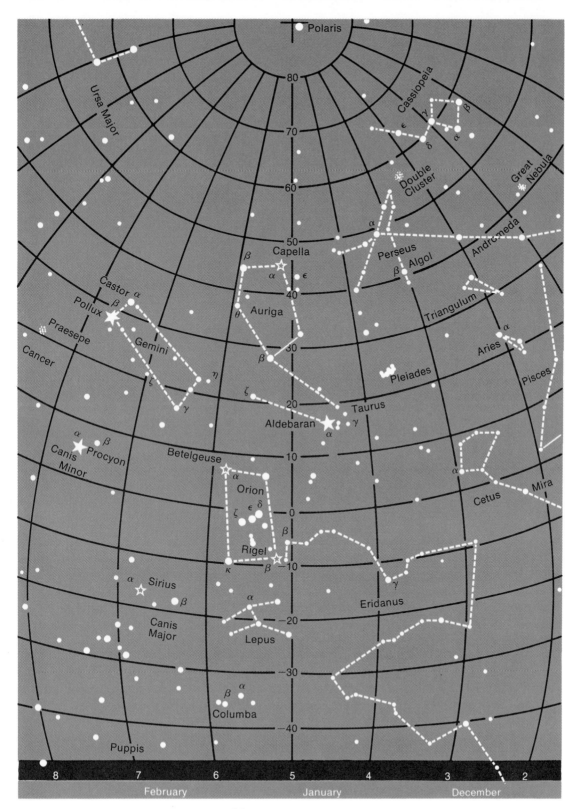

FIGURE 6.4.
Night sky in Winter (northern hemisphere).

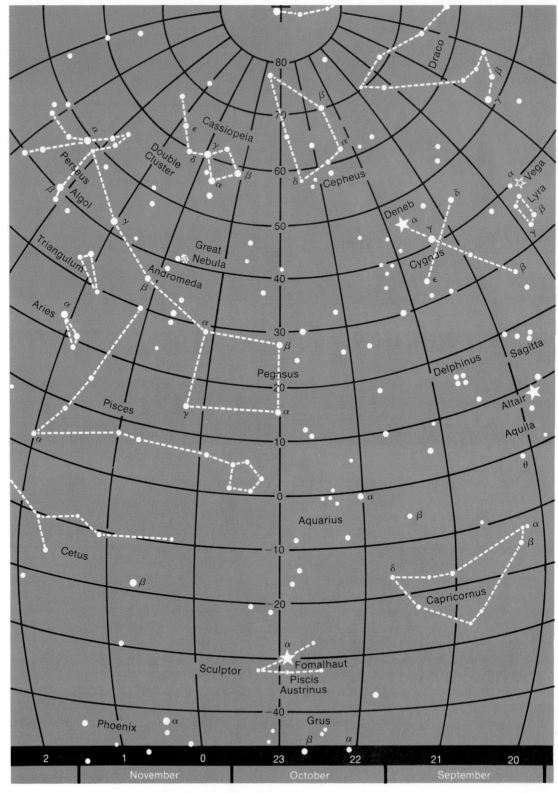

FIGURE 6.5.
Night sky in Autumn (northern hemisphere).

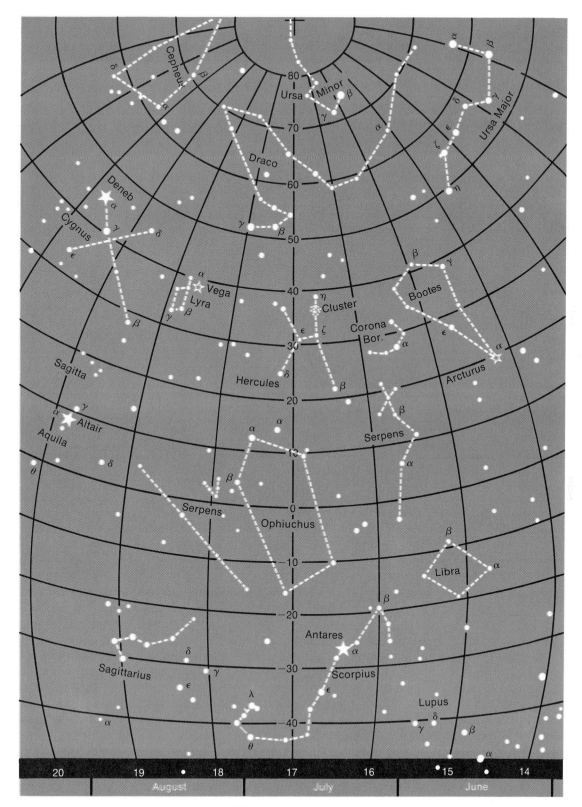

FIGURE 6.6.
Night sky in Summer (northern hemisphere).

TABLE 6.1.
*The constellations*

| Name | Approximate right ascension | Approximate declination | Intended meaning |
|---|---|---|---|
| Andromeda (And) | 01 | 35 | Andromeda |
| *Antlia (Ant) | 10 | −30 | Pump |
| *Apus (Aps) | 17 | −75 | Bird of Paradise |
| Aquarius (Aqr) | 22 | −15 | Water Bearer |
| Aquila (Aql) | 20 | 05 | Eagle |
| Ara (Ara) | 17 | −55 | Altar |
| Aries (Ari) | 02 | 20 | Ram |
| Auriga (Aur) | 05 | 40 | Charioteer |
| Boötes (Boo) | 15 | 30 | Herdsman |
| *Caelum (Cae) | 05 | −40 | Chisel |
| *Camelopardus (Cam) | 06 | 70 | Giraffe |
| Cancer (Cnc) | 09 | 20 | Crab |
| *Canes Venatici (CVn) | 13 | 40 | Hunting Dogs |
| Canis Major (CMa) | 07 | −25 | Big Dog |
| Canis Minor (CMi) | 07 | 05 | Small Dog |
| Capricornus (Cap) | 21 | −15 | Sea Goat |
| *Carina (Car) | 09 | −60 | Ship's Keel |
| Cassiopeia (Cas) | 01 | 60 | Cassiopeia |
| Centaurus (Cen) | 13 | −50 | Centaur |
| Cepheus (Cep) | 21 | 65 | Cepheus |
| Cetus (Cet) | 02 | −5 | Whale |
| *Chamaeleon (Cha) | 11 | −80 | Chameleon |
| *Circinus (Cir) | 16 | −65 | Compass |
| *Columba (Col) | 06 | −35 | Dove |
| *Coma Berenices (Com) | 13 | 20 | Berenice's Hair |
| Corona Austrina (CrA) | 19 | −40 | Southern Crown |
| Corona Borealis (CrB) | 16 | 30 | Northern Crown |
| Corvus (Crv) | 12 | −20 | Crow |
| Crater (Crt) | 11 | −15 | Cup |
| *Crux (Cru) | 12 | −60 | Southern Cross |
| Cygnus (Cyg) | 21 | 40 | Swan |
| Delphinus (Del) | 21 | 15 | Dolphin |
| *Dorado (Dor) | 05 | −60 | Swordfish |
| Draco (Dra) | 18 | 60 | Dragon |
| Equuleus (Equ) | 21 | 10 | Small Horse |
| Eridanus (Eri) | 03 | −25 | River Eridanus |
| *Fornax (For) | 03 | −30 | Furnace |
| Gemini (Gem) | 07 | 25 | Twins |
| *Grus (Gru) | 22 | −45 | Crane |
| Hercules (Her) | 17 | 30 | Hercules |
| *Horologium (Hor) | 03 | −55 | Clock |
| Hydra (Hya) | 10 | −15 | Water Monster |
| *Hydrus (Hyi) | 01 | −70 | Water Snake |
| *Indus (Ind) | 20 | −50 | Indian |

TABLE 6.1 (*continued*)

| Name | Approximate right ascension | Approximate declination | Intended meaning |
|---|---|---|---|
| *Lacerta (Lac) | 22 | 40 | Lizard |
| Leo (Leo) | 10 | 20 | Lion |
| *Leo Minor (LMi) | 10 | 35 | Small Lion |
| Lepus (Lep) | 05 | −20 | Hare |
| Libra (Lib) | 15 | −15 | Balance |
| Lupus (Lup) | 15 | −45 | Wolf |
| *Lynx (Lyn) | 09 | 40 | Lynx |
| Lyra (Lyr) | 19 | 35 | Harp |
| *Mensa (Men) | 06 | −75 | Table (Mountain) |
| *Microscopium (Mic) | 21 | −35 | Microscope |
| *Monoceros (Mon) | 07 | 00 | Unicorn |
| *Musca (Mus) | 13 | −70 | Fly |
| *Norma (Nor) | 16 | −55 | Square |
| *Octans (Oct) | 22 | −85 | Octant |
| Ophiuchus (Oph) | 17 | 0 | Snake Bearer |
| Orion (Ori) | 05 | 00 | Orion |
| *Pavo (Pav) | 20 | −60 | Peacock |
| Pegasus (Peg) | 22 | 20 | Pegasus |
| Perseus (Per) | 03 | 40 | Perseus |
| *Phoenix (Phe) | 01 | −45 | Phoenix |
| *Pictor (Pic) | 07 | −60 | Easel |
| Pisces (Psc) | 00 | 10 | Fishes |
| Piscis Austrinus (PsA) | 23 | −30 | Southern Fish |
| *Puppis (Pup) | 07 | −35 | Ship's Stern |
| *Pyxis (Pyx) | 09 | −35 | Ship's Compass |
| *Reticulum (Ret) | 04 | −65 | Net |
| Sagitta (Sge) | 20 | 15 | Arrow |
| Sagittarius (Sgr) | 18 | −30 | Archer |
| Scorpius (Sco) | 17 | −35 | Scorpion |
| *Sculptor (Scl) | 01 | −30 | Sculptor |
| *Scutum (Sct) | 19 | −10 | Shield |
| Serpens (Ser) | 16 | 05 | Snake |
| *Sextans (Sex) | 10 | 00 | Sextant |
| Taurus (Tau) | 05 | 20 | Bull |
| *Telescopium (Tel) | 18 | −45 | Telescope |
| Triangulum (Tri) | 02 | 35 | Triangle |
| *Triangulum Australe (TrA) | 16 | −65 | Southern Triangle |
| *Tucana (Tuc) | 23 | −60 | Toucan |
| Ursa Major (UMa) | 11 | 50 | Great Bear |
| Ursa Minor (UMi) | 15 | 75 | Small Bear |
| *Vela (Vel) | 09 | −50 | Ship's Sails |
| Virgo (Vir) | 13 | 00 | Virgin |
| *Volans (Vol) | 08 | −70 | Flying Fish |
| *Vulpecula (Vul) | 20 | 25 | Fox |

*Of modern origin.

FIGURE 6.7.
The Orion Nebula, a cloud of gas in which stars are now being born. (Courtesy of the Hale Observatories.)

The names of the constellations developed from association of star patterns with quaint terrestrial images. These names have no scientific significance, of course, but there is a continuing attraction in many of these fanciful descriptions. The numbers given in Table 6.1 represent an astronomical system of latitude (declination) and longitude (right ascension). These numbers can be used to locate the different constellations in the star maps of Figures 6.1 to 6.6.

It is a curious thought that modern man is probably less familiar with the night sky than were our remote ancestors of 100,000 years ago. Life in cities tends to change our mode of existence to a state in which we are hardly aware of the sky, or indeed of many other aspects of our natural environment. Yet I doubt whether even the most hard-bitten soul can look up at the sky on a clear, dark night without wondering where it all came from, and what it all means. What it *means* is a difficult and so-far unsolved problem, but at least we can say something about where the stars come from. Stars are forming right now within the clouds of gas, like the Orion Nebula, shown in Figure 6.7. This cloud can easily be seen in the "sword" of the constellation of Orion—a wide-angle photograph of Orion is shown in Figure 6.8. The diameter of the Orion Nebula is about 15 light years, and its distance from us is about 1,500 light years.

A light year is simply the distance traveled by

FIGURE 6.8.
A wide-angle photograph of the constellation of Orion.
(Courtesy of the Hale Observatories.)

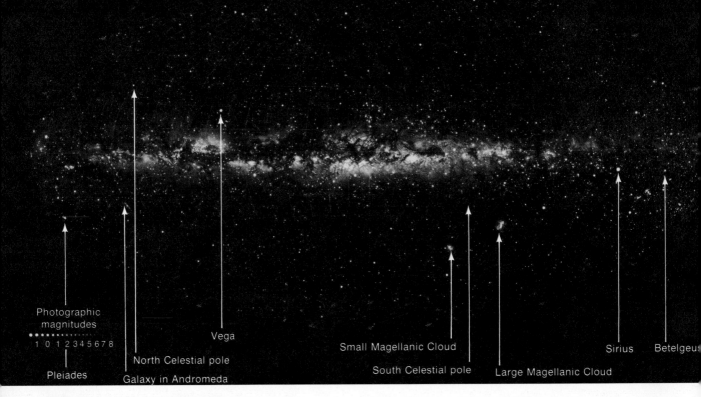

Photographic
magnitudes

1 0 1 2 3 4 5 6 7 8

Pleiades

Galaxy in Andromeda

North Celestial pole

Vega

Small Magellanic Cloud

South Celestial pole

Large Magellanic Cloud

Sirius

Betelgeus

FIGURE 6.9.
A complete representation of the whole Milky Way, built up from many photographs.
(Courtesy of the Observatorium, Lund, Sweden.)

light in a year. Since light has a speed of 300,000 kilometers per second, and since there are rather more than 30 million seconds in a year, it follows that light travels nearly 10 million million kilometers in a year. The distance of the Orion Nebula from us is thus about 15,000 million million kilometers.

Even so vast a distance as this is still not much more than 1 per cent of the over-all scale of the atar system in which we live, usually known as the galaxy—although the nearer part of the galaxy is often referred to colloquially as the Milky Way. A reconstruction of how the Milky Way looks when observed in the sky is shown in the form of a map in Figure 6.9.

There are many gas clouds like the Orion Nebula to be found along the Milky Way; the Orion Nebula is simply one of the nearest. At distances up to 5,000 light years, gas clouds such as the Rosette Nebula (shown in Figure 6.10), are found. However, we cannot see the whole of our galaxy, because a myriad of tiny particles (even smaller than cometary dust) exists everywhere throughout the Milky Way. These particles act like a kind of fog, which absorbs light after it has traveled for more than a few thousand light years, so that the full scale of our galaxy, encompassing distances of about 100,000 light years, is not seen in Figure 6.9.

Sometimes the obscuring particles are highly concentrated into what are called *dark nebulae*,

FIGURE 6.10.
The Rosette Nebula. (Courtesy of the Hale Observatories.)

FIGURE 6.11.
The Horsehead Nebula, a dark nebula produced by myriads of fine dust grains.
(Courtesy of the Hale Observatories.)

of which the Horsehead Nebula (shown in Figure 6.11) is a striking example. Here the fog is so thick that we see nothing on the far side of the nebula—it blots out everything lying behind it. In contrast to the dark nebulae, the *bright nebulae,* like Orion, are made to shine by the stars which have formed within them. Light emitted by the stars is first absorbed and then reradiated by atoms of gas within the nebula. The general order of density in clouds like Orion is about 10,000 atoms per cubic inch, but much higher gas densities are found in smaller clouds. There are such regions of especially high concentration within the Orion Nebula itself,

and it is within these regions that stars are being formed.

The total mass of gas in a cloud like Orion is sufficient to form a very large number of stars, perhaps as many as 100,000 stars. Thus when regions of particularly high density fragment themselves into stars, they form not just one star, but a whole cluster of stars. A very large cluster is shown in Figure 6.12. However, a cluster that is more typical of star formation now is shown in Figure 6.13. Clusters like Figure 6.12 were formed in the early history of our galaxy, at a time when our galaxy is believed to have been itself a vast con-

FIGURE 6.12.
A globular cluster containing perhaps 100,000 stars.
(Courtesy of the Hale Observatories.)

FIGURE 6.13.
The bright stars of the Pleiades cluster are blue and highly luminous.
(Courtesy of the Hale Observatories.)

densing cloud containing sufficient gas to make about 100,000 million stars. The cloud is thought to have fallen together to form a disk with the kind of structure shown in Figure 6.14, a flat pancake but with a developing bulge toward the center. The Sun and the planets of our solar system lie far out from the center, and even the stars we see in the sky—the stars which form the constellations—are all comparatively close to us, lying in the small region marked in Figure 6.14.

Let us think about the condensation of a single star in a little more detail. Think of a ball of gas, initially with a diameter much bigger than a star,

gradually shrinking and becoming more and more dense as it does so. Now whenever a gas becomes compressed—whenever it becomes more dense—its temperature tends to rise. The compression of air in a bicycle pump produces heat—the barrel of the pump becomes warm whenever the pump is used vigorously, because the barrel picks up heat from the gas which is being compressed. So we expect the interior of our protostar to become hotter and hotter as it shrinks. Gas near the center is more compressed, and becomes hotter than gas near the surface. This difference of temperature causes heat to flow from the central regions to the surface. The

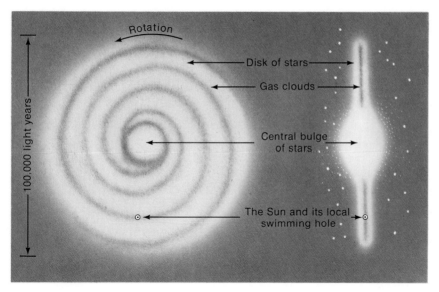

FIGURE 6.14.
A schematic drawing of the Milky Way, see face-on in (a) and edge-on in (b).
Most of the stars we see by naked eye belong to a comparatively small region
immediately surrounding our solar system.

energy thus arriving at the surface is then radiated away into space: the protostar begins to shine, and calculation shows that it does so with an orange-red color.

The process goes on until the central regions become so hot that nuclear processes which convert atoms of hydrogen into atoms of helium become important. Eventually the conversion becomes rapid enough that all the energy which flows out from the center, and is then radiated into space from the surface, is supplied by the nuclear processes. The star then comes into a steady situation, without any further shrinkage being necessary to make good the energy losses. It is at this stage, when the star has become a stable nuclear reactor, that we speak of the star as having "formed." It is no longer a "protostar."

It is natural to expect that different stars will form out of different quantities of gas, that their masses will be different one from another. And it is natural to expect those of larger mass to be brighter than those of small mass. What is perhaps not so expected is the strength of this effect. Stars of large mass are very much brighter than those of small mass. At the time of their formation, stars of large mass are violet-blue in color, whereas stars of small mass are a dark red.

Stars of large mass die rather quickly, because they quickly use up their supply of hydrogen atoms. When the supply of hydrogen becomes exhausted, no more helium can be made by the nuclear reactions. A star like the Sun has a "life" of about 10,000 million years, whereas the stars of large mass—fifty times the mass of the Sun—live for only a few million years. It follows that stars of large mass are only seen among groups that formed

FIGURE 6.15.
The star cluster known as M67. (Courtesy of the Hale Observatories.)

recently. In an old group of stars, like that of Figure 6.15, the stars of large mass died long ago. Such old groups of stars have a red color, since the stars still shining within them, being stars of small mass, have a red color. Young groups of stars, on the other hand, are blue in color, because they still contain stars of large mass, which are blue and which dominate the light from the group.

There are many galaxies other than our own, of which the nearest large one is shown in Figure 6.16. This is the Andromeda Nebula, whose position on the sky in the constellation of Andromeda can be found in Figure 6.5. The central region is seen to be orange-yellow, whereas the outermost regions are rather blue, which means that the stars of the central region are old. Young stars in this galaxy occur, however, in the outer regions, not in very great number, but in a sufficient number to dominate the light of the fainter outer regions. The brighter central region of Figure 6.16 is visible to the naked eye and can be found on a clear dark night (away from city lights!) with the help of Figure 6.5.

How are we to reconcile the idea that the stars are all born in clusters with the fact that most stars are not today to be found in clusters? The

FIGURE 6.16.
The galaxy known as the Andromeda Nebula, in color. The difference between the inner yellow region and the blue of the outer regions can be explained in terms of the ages of the stars. (Courtesy of the Hale Observatories.)

answer to this question is that the gravitational pull of the whole of our galaxy causes clusters, especially the smaller clusters, to disintegrate and break up. Large clusters, like that in Figure 6.12, do tend to survive, but such large clusters are rather rare. Most stars were probably born in smaller clusters which by now have been broken apart.

Yet stars, even after they become free, retain a kind of "memory" of their cluster origin. This memory shows itself in the fact that a high proportion of stars belong to "binary" systems, which are star pairs in which one star moves in an orbit around the other star. Fortunately for us, our Sun does not belong to such a pair; otherwise the Earth would be subject to the gravitational influence of two stars. This influence would be so variable that a stable orbit for the Earth would quite likely not be possible. We should be subject to such extremes of hot and cold that life here would be hazardous and perhaps not even possible at all.

Some systems contain more than two stars. Castor, one of the two bright stars in the constellation of Gemini, when seen in a telescope under good conditions, appears as a triple system. The two brighter components of this triple system move around each other in a period of rather more than 300 years. The third, fainter companion moves around the other two in a time that exceeds 10,000 years. Furthermore, each of the three stars of this triple system has turned out to be a closely spaced double, not detectable directly by eye but only by instrumental means. Castor is thus a system containing six stars. Quite likely it is a remnant of an old cluster. Life in such a system, if it were possible, would certainly not lack for variety.

What happens to stars which use up all their supplies of hydrogen? How do stars die? Hydrogen

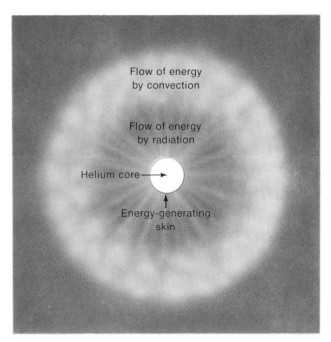

FIGURES 6.17.
The structure of a star as it enters the giant phase of its life.

becomes exhausted first in the central regions of a star. The central regions then begin to shrink, becoming hotter as they do so. Hydrogen still remains in regions remote from the center, and this non-central hydrogen still remains to be converted into helium. At first sight we might also expect the outer regions of the star to shrink, but calculation has shown just the reverse: the outer part of the star expands, and in many cases it does so enormously. The star is now said to have become a "giant." The color of the light radiated into space from the surface becomes very red. The general form of this giant structure is illustrated in Figure 6.17. The star Betelgeuse in the constellation of Orion is an example of such a giant. Indeed, Betelgeuse has become so distended in its structure that, if it were to replace the Sun, it would extend past the orbits of the four inner planets, Mercury, Venus, Earth, and Mars, as in Figure 6.18.

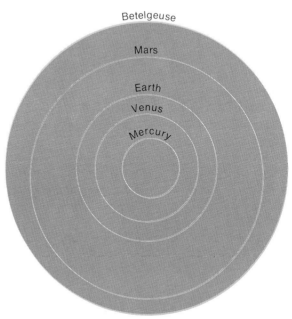

FIGURE 6.18.
The star Betelgeuse in the constellation of Orion is so large that, if it replaced the Sun, all the inner planets would lie within it.

Maximum      Minimum      Maximum

FIGURE 6.19.
The size variation of a pulsating star.

Stars which become as large as this tend to develop what are known as pulsations. They vary in diameter cyclically, as is illustrated in Figure 6.19. The light emitted at maximum diameter is redder than at minimum diameter. Since very deep red light tends to become invisible to the eye—as one notices when a visibly red hotplate is allowed to cool—it follows that such a pulsating giant star can become invisible at maximum diameter. This is the situation for the star Mira. As it approaches minimum diameter, Mira is a quite bright red star, easily seen by naked eye. As it approaches maximum size, the light from Mira becomes so very red that it cannot be seen at all by the naked eye. To early astronomers, Mira was an astonishing phenomenon, a star appearing regularly in the sky every 11 months, and then disappearing again! The Arabic name means "The Wonderful." The fact that Mira was so well known to early astronomers shows the care with which they watched the sky. No doubt its nearly annual appearance seemed

fraught with great significance. But of what? Astronomical soothsayers, equipped with a fair ration of commonsense, could no doubt use the superstition generated by the appearance of Mira as a device for impressing their views on the rulers of the day, much as economists have succeeded in impressing their views on present-day governments.

Many nuclear processes besides those which convert hydrogen to helium are possible. What is special about the reactions which convert hydrogen to helium is that they are the ones which operate at the lowest temparatures. Consequently they are the first ones to become important as the temperature rises inside a condensing protostar. The further contraction of the star which sets in as the hydrogen becomes exhausted raises the temperature further, and this brings other reactions into operation. The helium itself can undergo reactions which have the important effect of producing the elements carbon and oxygen, so important to life on the Earth. At still higher temperatures, carbon and oxygen can themselves undergo further reactions which generate a whole host of well-known elements: sodium, magnesium, aluminum, silicon, sulfur, and calcium. At even higher temperatures, the common metals iron, nickel, chromium, manganese, and cobalt appear. Yet all these processes, very complex in detail, can provide a star with only a limited quantity of energy. All these reactions are like a limited number of checks that the star can cash. Once they are cashed, no more energy from nuclear processes can be forthcoming. What can the star do then?

Perhaps cool off, becoming a dead star which ceases to shine? Calculation shows that a star cannot "die" in this way unless the mass of the star is less than a certain upper limit, which turns out to be not too different from the mass of the Sun. In order to be able to cool off, a star more massive than this must fling material out into space—from where all its material originally came. There are two main ways in which this shedding of excess material seems to happen.

Stars of moderate mass shed most of their unwanted material rather gently, probably ending with a fling in which a cloud known as a *planetary nebula* is thrown off. Examples of planetary nebulae are shown in Figures 6.20 and 6.21. The residue left behind eventually cools to become a remarkable kind of star known as a *white dwarf*. White dwarfs, although they have masses quite comparable to that of the Sun, have diameters comparable to those of the inner planets, which implies exceedingly high densities within them. The material at the center of a white dwarf can have a density so high that a volume equal to that of an ordinary sugar lump contains a ton of material.

Stars of large mass are still more violent in their behavior. They become unstable, and explode like a nuclear bomb. For a few days following such an explosion, usually referred to as a *supernova,* the

FIGURE 6.20
A planetary nebula NGC 7293. (Courtesy of the Hale Observatories.)

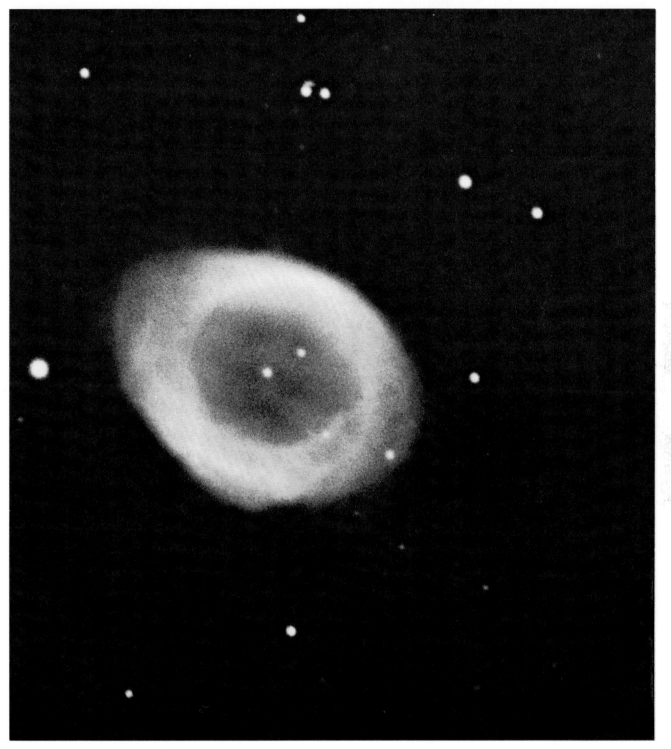

FIGURE 6.21.
The Ring Nebula NGC 6720, another planetary nebula.
(Courtesy of the Hale Observatories.)

star is temporarily as bright as the whole of the Milky Way. Such explosions occur in our galaxy with a frequency of about one per century. A supernova observed by Chinese astronomers in the year A.D. 1054 produced the object shown in Figure 6.22. From its appearance this object was named (long before its nature was understood) the Crab Nebula. The supernova was seen by Chinese astronomers on July 4, 1054, as a "guest star visible by day like Venus." It remained as bright as Venus until July 27. Thereafter it faded gradually, becoming invisible to the naked eye by April 17, 1056.

It is curious that no similar accounts of this supernova have been found in European or Arabic records. An apparition in the sky at least as bright as Venus can hardly have been missed. It remained exceedingly bright for about a month, and it is hard to believe that European, North African, and Arabian skies could have been cloud-covered for as long as this. Rather must one suppose that the supernova was seen by millions of people. Most would be illiterate and therefore unable to record it. What the specialized groups of chroniclers chose to record seems to have turned on their beliefs. In China, it was thought that terrestrial events could be foretold from occurrences in the sky, and therefore peculiarities in the sky were avidly searched for, and noted. In Europe, on the other hand, the monkish chroniclers believed the heavens to be the completed handiwork of "God," and therefore to be perfect and not subject to change. Any record to the contrary would have been heretical and would have provoked an angry response from theologians and philosophers, just as Galileo's discovery of spots on the Sun did some five centuries later.

FIGURE 6.22.
The Crab Nebula. The light from this object is generated by electrons of very high speed, moving in a magnetic field. (Courtesy of the Hale Observatories.)

FIGURE 6.23
Two Pueblo Indian pictographs, on the left from Navajo Canyon, on the right from White Mesa. (Courtesy of William C. Miller.)

Two pictographs, shown in Figure 6.23, have been reported by William C. Miller. They were discovered at different sites in the cave dwellings of the Pueblo culture of the North American Indians, a culture which spanned the year A.D. 1054. Miller believes that the two pictographs refer to a moment when the Moon in its monthly motion across the sky approached close to the "guest star" of the Chinese. He rejects the possibility that the pictographs refer to a juxtaposition of Venus and the Moon on the grounds that Venus and the Moon approach each other closely every few years, whereas the fact that no other astronomically oriented pictograph has so far been discovered suggests that Figure 6.23 refers to a unique and most remarkable event. So it may be that,

whereas the literate men of Europe were too prejudiced to note the event, the non-literate Indians of North America nevertheless managed to leave a record behind them. The pictographs were chipped in rock with the artist's back to the sky. Looking over his shoulder, would he denote the crescent moon with its actual left-right sense, or would he invert it, as if he were turned full-faced to the sky? These violently exploding massive stars also leave a residue behind, a residue even more remarkable than a white dwarf. The core of such massive stars becomes too dense for a white dwarf, attaining a fantastic condition in which a cube of the size of a sugar lump contains 100 million tons of material. They become objects known as *neutron stars*. Although they have masses comparable to

White dwarf  Neutron star                Earth

that of the Sun itself, neutron stars are much smaller in size than the planets, as can be seen from Figure 6.24. Indeed, neutron stars are not far removed from the condition which astronomers and physicists refer to nowadays as a *black hole,* a condition where matter collapses in on itself and ultimately disappears altogether from our universe.

Because of their small diameters, the neutron stars spin around very quickly, most of them about once a second, although exceptional ones may spin even faster. As they spin, a lighthouse effect, illustrated in Figure 6.25, seems to operate, in which all kinds of radiation—radiowaves as well as visible light, and in some cases even x-rays—sweep periodically across an external observer. When this happens, the star is known as a *pulsar.* There is a

FIGURE 6.25.
Pulsars are thought to behave like the rotating beam of a lighthouse.

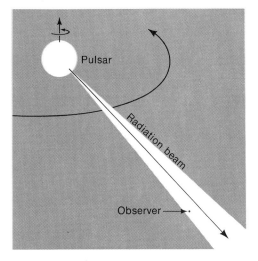

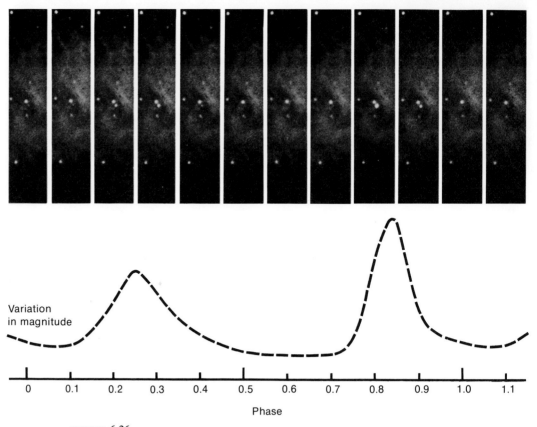

FIGURE 6.26.
A variable star-like object in the Crab Nebula is believed to be a rotating neutron star. (Courtesy of the Kitt Peak National Observatory.)

pulsar within the Crab Nebula. It turns very rapidly, at about thirty rotations per second, and as it does so the light from it goes on and off, as can be seen from the series of photographs shown in Figure 6.26. This indeed is some lighthouse.

Our galaxy is believed by astronomers to have been composed initially of hydrogen and helium only. The evidence for this belief is that old stars contain smaller quantities of the other elements than young stars do, elements like carbon, oxygen, sodium, magnesium, silicon, sulfur, calcium, iron, chromium, manganese, nickel, copper, and zinc. These and many other elements are thought to have been produced by the nuclear processes within stars. Subsequent to their production inside stars, all these materials are broadcast into space by stellar explosions. We thus have the cyclical picture of Figure 6.27, in which material is transferred back and forth between stars and gas clouds like the Orion Nebula, where it is available to become a part of new stars when they form.

The implication of this idea is that the common materials of our daily world, the carbon that is the basis of life, the oxygen we breathe, the metals we use, have all experienced the cycle of Figure 6.27. They were produced in stellar furnaces, and

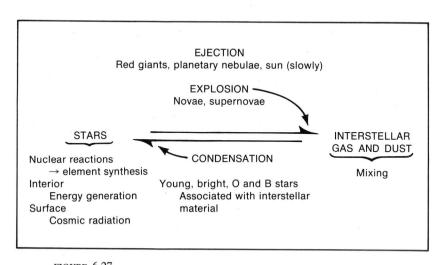

FIGURE 6.27.
A cyclical process, in which matter is transferred backward and forward between stars and the interstellar gas.

they had been flung into space before the Sun and the planets of our system were formed. The constituents of our daily world have not been fashioned here "on site," within our solar system. Rather, they were fashioned many aeons ago within other stars that are by now faint white dwarfs or super-dense neutron stars which we have no means of identifying. Voltaire was right when he wrote of "sermons in stones," but for a reason he could not have appreciated. The silicon, magnesium, aluminum, and oxygen making up a commonplace piece of stone, have a remarkable, and even fantastic, history behind them.

# 7

## *Life in the Universe*

# 7

*Life in the Universe*

There may be forms of life that we are totally unaware of, forms that we can barely conceive. One might speculate, for example, on the possibility of life existing at very high densities, 100 million tons to a sugar lump cube. However, given what we know at present about life, we have no way to undertake serious discussions of such speculations. In this chapter we shall bypass all such strange possibilities, keeping to issues relating closely to life as we know it.

We believe we know today quite a bit about our kind of life—whereas fifty years ago we did not. We know quite a bit about the way life reproduces itself from one generation to another, about the whole chemical structure that makes such reproductions possible. In terms of this knowledge, it is interesting to consider what we can say about the emergence of life and of intelligence in the universe.

Life appears to have formed on the Earth out of quite simple substances—water, carbon dioxide,

ammonia, hydrogen cyanide (an association formed by an atom of carbon, an atom of hydrogen, and an atom of nitrogen). Recently astronomers have found just these substances to be present in very large quantities within gas clouds like the Orion Nebula (Figure 6.7) and like the clouds shown in Figures 7.1 and 7.2. It has thus become clear that the simple, basic life-forming molecules exist everywhere in our galaxy. Since the chemical processes which led to the emergence of life on the Earth are the same everywhere else, it is reasonable to suppose that life must also have arisen on many of the planetary systems which move around other stars in our galaxy. There are 100,000 million stars in our galaxy, and a large fraction of them are of a type which, astronomers believe, should possess planetary systems. So in considering the emergence of life on a galactic scale, we have something on the order of 100,000 million possible planetary sites to consider.

Of course, not all planets are suitable for life.

FIGURE 7.1.
The Trifid Nebula. (Courtesy of the Hale Observatories.)

FIGURE 7.2.
The Lagoon Nebula in the constellation of Sagittarius.
(Courtesy of the Hale Observatories.)

In our own system the Earth is quite different from the other planets, as we can see from comparing Figure 7.3 with the sterile surface of the Moon in Figure 7.4. If the Sun had possessed eight planets instead of nine, without the Earth, the solar system would very likely have been sterile. Among other systems of planets moving around stars, many may well be sterile. Yet with so many possibilities available, this situation need scarcely worry us. Rather we should ask: Having allowed for all the unfavorable cases, does a large number of suitable sites still remain? If only 10 per cent of all systems were to be astronomically favorable, and if only a further 10 per cent of these had the right chemical materials, we would still have as many as 1,000 million favorable possibilities.

The first step toward the origin of life is comparatively well understood. In this step, the simple chemical materials mentioned above form more complicated associations, like sugars and amino acids. The essential feature of this first step is that it stores energy, which can then be used to drive more complex systems. The source of this energy must be the light which shines on the planet, the light from the central star. This first step is not so complex that it is unlikely to happen; most chemists and biologists seem to have little doubt that it would happen in most cases.

In the second step, materials like sugars and amino acids link up into the exceedingly intricate chemical structures on which life itself is based. Much work is going on today, seeking to discover how the first biological cell—the first cell able to reproduce itself—came into being. Until more is known about this second step, we cannot estimate the chance that life will emerge in those systems, perhaps 1,000 million of them, where the astronomical and chemical conditions are favorable. It *may* be that the emergence of life depends on some very improbable circumstances. However, in our study of the astronomical and chemical conditions needed for life to emerge, the pattern so far has been that life becomes more and more probable in the universe as we learn more and more about details: what at first sight seems incredible and unattainable becomes credible and attainable as our knowledge increases. Given this pattern, it does not seem unreasonable to suggest that life has actually arisen in many millions of our 1,000 million favorable sites. The likely distance from us of the nearest planetary system on which life has arisen would then be about 100 light years, well within the range to which stars can be seen with the naked eye.

The third step in this emergence of life has been considered already in Chapter 1. We saw in Figures

FIGURE 7.3.
The Earth as seen on the Apollo 16 mission. (Courtesy of NASA.)

FIGURE 7.4.
The sterile surface of the Moon. (Courtesy of NASA.)

1.8 and 1.9 how more and more complex life forms developed on the Earth. It is interesting to contemplate how rich may be the variety of creatures which have arisen on all favorable sites throughout the entire galaxy. In how many of these cases may intelligent creatures have emerged?

Our emotional attitude to life is not really a chemical matter at all. Although the difference between a well-loved person's being alive and being dead may turn on certain subtle chemical processes, this is not at all the way we feel about it. Most people who would never dream of strangling a dog do not hesitate to swat a mosquito. Yet the chemistry of the mosquito is basically the same as that of the dog. The situation is that we distinguish between "higher" and "lower" animals according to the complexities of the nervous systems with which the animals are endowed. A nervous system is basically electrical in its operation, with an animal being made up of a chemical system plus an electronic one.

$$\text{Animal} \equiv \text{Chemical replication} + \text{Electronic system.}$$

The more the electronic part dominates this summation, the "higher" we judge the animal to be in the zoological evolutionary scale. The more the electronic system happens to match our own system, the more favorably we regard the animal. And among humans, the more similar the other person's electronic system is to our own, the better regarded, or the better loved, the person is. Similarity, or otherwise, distinguishes the category of "us" from the category of "them."

At a certain level of electronic complexity, we rather arbitrarily introduce the notion of "intelligence," a level set a little below our own capacity. We acknowledge a spark of intelligence in the dog, but the behavior of a cat strikes many people as "independent" rather than intelligent. Essentially as a matter of definition, any creature with an electronic system more complex than our own would be endowed with high "intelligence."

Animals are not really able to synthesize amino acids and sugars, as plants do. Animals must therefore acquire these substances either by eating plants or by eating other animals. Basically, all animals are scroungers, living on the stored chemical potentialities which others have first accumulated. It was precisely to assist in the process of scrounging that the electronic systems possessed by animals developed. Since the better the electronic system the better the scrounger, biological evolution has operated steadily during many millions of years to increase the complexity of animal electronics. And since we judge the level of an animal by the complexity of its electronics,

it follows that the higher the animal, the greater the scrounger—with man himself sitting at the top of the pyramid.

From the foregoing consideration of the nature of an "animal," it therefore seems clear that an electronic (nervous) system would very likely develop for all animals, everywhere. That is, the need to search for food would make "eyes" a normal development. Animals with eyes are then likely to prey on each other, with biological evolution forcing the development of "weapon systems"—claws, teeth, and ultimately a thinking brain, the most deadly weapon of all. The logical sequence leading to the emergence of a thinking brain appears inevitable; so we can expect this sequence to have happened quite generally.

The electronic system in man has indeed become so subtle that our scrounging for energy, in particular, has now extended well beyond the eating of plants and of other animals. We scrounge extensively today on nonliving materials. The discovery of fire made use of trees as an energy source. The burning of coal and oil were further steps along the same path, and in the modern nuclear power plant we have attained to the use of entirely nonorganic materials as an energy source. This access to nonanimal sources of energy has developed with increasing rapidity, to a point in our modern society where we can clearly see that either some

more restrained pattern of behavior must be applied in future years or the evolution of our species will end itself in a catastrophic social explosion. It is in these evidently crucial circumstances that we have begun to wonder how things may have fared with other creatures living on planets moving around other stars, and we have even begun to wonder about the possibility of communicating with such creatures.

*Interstellar communication*, as we may call it, raises many problems, some technical, some of general interest. Let it be said immediately that the only feasible mode of communication between creatures living on different planets moving around different stars would seem to be by radio. A vast array containing 900 individual radiotelescopes, each with a diameter of 100 meters—similar to the largest fully steerable radiotelescope yet constructed, shown in Figure 7.5—has been proposed.

FIGURE 7.6
A schematic drawing of the massed radiotelescopes of *Project Cyclops.*
(Courtesy of NASA.)

Such an array could actually be built now, and would be able to achieve interstellar communication. It has been named *Project Cyclops,* and is shown schematically in Figure 7.6.

It is worth noticing that physical travel by us to distant stars is probably not possible and even if it were, physical travel would take much more time than an interchange of messages by means of a system like *Project Cyclops.* The impossibility of physical travel to distant stars might seem at first sight to imply a decline in romantic possibility, a loss of richness in the scheme of things. But a little thought soon shows that precisely the opposite is true. If physical travel from one planetary system to another were feasible, then the first creatures to

become technologically capable of space travel would be likely to spread themselves everywhere throughout the galaxy—just as science-fiction writers are always imagining the human species to do. It would be only too likely that the galaxy would thus come to have only one form of intelligent creature. This indeed would be a loss of richness. But with space travel *not* possible, creatures in one planetary system cannot interfere with the physical development of creatures in other systems. Many possibilities, with great potential richness are then permitted.

We come now to what appears to be the most uncertain question of all. Given a suitable planet, given the origin of life, given the emergence of intelligence to a level at least equal to our own, for how long on the average can we expect such an intelligence to persist? Even if intelligence arises on as many as a million planets in our galaxy, there will still be very few intelligent species around *at the present moment* unless high intelligence, once it arises, persists for more than 10,000 years, for the following reason. The age of our galaxy, the time span throughout which intelligence can emerge, is very long indeed—about 10,000 *million* years. Unless intelligence lasts once it arises, there will be very little overlap in time between its brief emergence on one planet and its emergence on another planet.

Suppose that our capacity to build an instrument of the technological quality of *Project Cyclops* might last for only 10,000 years. Is this an overly pessimistic assessment of the future of the human species? In view of the state of our present-day society, is it not rather an optimistic assessment? When one contemplates the huge human populations that have grown with startling suddenness during the last century or so, when one contemplates the excessive modern pressure on natural resources, it is hard to summon much confidence in a future extending more than a few decades. Devastating crises, one feels, must overtake the human species within a hundred years at most. We are living today, not on the brink of social disaster, as we often tend to think, but actually *within* the disaster itself. This is exactly what the news media report to us every day.

We have seen that the phenomenon of "intelligence" is an outcome of aggressive competition. Intelligence and aggressiveness are coupled together inevitably by the mechanisms of biological evolution. An intelligent animal anywhere in the galaxy must necessarily be an aggressive animal and must necessarily become faced at some stage by the same kind of social situation as that which now confronts the human species. Inevitably, then, "intelligence" contains within itself the seeds of its own destruction. Can any solution be found for this inherent difficulty? We can approach this critical question by considering what would be needed here on Earth for our species to be able to maintain itself at a high technological level for a period many times longer than ten thousand years.

There was a time, not many centuries ago, when

the total human population of the whole Earth was no more than 500 million. The energy from the combustion of wood would then have been sufficient to provide such a population with a high standard of life. The tragedy of our species is that, when the population was that low, the technology necessary for the effective industrial use of sunlight was not available. By now the technology is available, but unfortunately the human population has now risen to about four billion, and at this higher number the simply acquired energy of sunlight is not adequate. The very large present-day human population therefore demands energy sources of a short-term character, particularly from the consumption of coal and oil. Since these short-term sources cannot last, we are faced by the exceedingly difficult problem of producing nuclear energy in large amounts. Eventually, the technical difficulties of producing nuclear energy may force us to contemplate a reduction of the total human population back to a level of 500 million. If this were done in a controlled way, say, at a rate of 1 per cent per annum, it would take about two hundred years to achieve, which is about how long the world's coal reserves may be expected to last.

Thereafter, using sunlight, adequate energy for 500 million people would be available for a future of essentially unlimited duration.

Such is the crisis which faces us during the next several centuries, and which may prove to be as crucial as any encountered in the long chain shown in Figures 1.8 and 1.9. Its resolution might indeed be taken to herald the appearance of a new species on the Earth. Among our millions of intelligent species in our galaxy, many have likely passed already through a similar crisis. It is among these that we may expect interstellar communication to be taking place, perhaps by means of systems like that of Figure 7.6. For creatures who can foresee a long-term stable future of millions of years ahead of them, the interval of several hundred years needed between the transmission of a message and the reception of a reply to it would not be a serious impediment. There would be ample time for many messages to be interchanged. No species with real confidence in its future would hesitate to search the galaxy for the other intelligences which must surely have emerged and which may have triumphed over the difficulties that still confront the human species today.

# 8

## *Galaxies and the Universe*

# 8

## *Galaxies and the Universe*

Our galaxy is a member of a "local group" of about twenty galaxies. The only other large member of this group is the Andromeda Nebula, whose spiral pattern can be seen clearly in Figure 6.16. Another spiral galaxy in this local group is shown in Figure 8.1. Both this galaxy, usually described by the catalogue number M33, and the Andromeda Nebula are distant from our own galaxy by about 2 million light years.

The remaining members of this local group are small galaxies, often referred to as "dwarfs." Two of them happen to be comparatively close to our galaxy, at a distance of about 150,000 light years. These are the Large Magellanic Cloud and the Small Magellanic Cloud, shown respectively in Figures 8.2 and 8.3. These clouds are extensively studied by astronomers because, being comparatively close to us, they reveal more detail than can be seen in any other galaxy external to our own.

Unfortunately, they are located toward the south pole of the sky, and so cannot be seen at all from observatories in the northern hemisphere of the Earth. Their precise positions can be found on the star map of Figure 6.2, and the galaxy M33 can be located in Figure 6.5. Light from the spiral arms of M33 comes mainly from a comparatively small number of very bright stars. These are stars of exceptionally high mass which have formed recently from the gas and dust within this galaxy. The presence of dust in Figure 8.1 is easily seen.

We look at the galaxy M33 more or less face on, whereas we look at the Andromeda Nebula much more sideways on, which is why the Andromeda Nebula appears to us to have an oval shape.* The stars which can be seen in both Figures 6.16 and

*The orientation is easily judged by drawing a circle on a sheet of paper and by then turning the paper so as to produce a similar oval shape.

FIGURE 8.1.
The Sc galaxy M33. (Courtesy of the Hale Observatories.)

FIGURE 8.2.
The Large Magellanic Cloud. (Courtesy of Dr. V. M. Blanco,
Cerro Tololo Inter-American Observatory.)

FIGURE 8.3.
The Small Magellanic Cloud. (Courtesy of Dr. V. C. Reddish
and the Science Research Council.)

8.1 are simply foreground stars of our own galaxy; we have to look out through our own galaxy in order to see the systems which lie outside.

When we look out to distances of 10 million light years or more, galaxies are found in increasing numbers. Examples are shown in Figures 8.4 to 8.8, which show further cases of remarkable spiral structure. All these systems are disclike in form. In Figures 8.4 and 8.5 we are looking at the disc structure nearly face on, in Figures 8.6 and 8.7 at intermediate angles, and in Figure 8.8 sideways on.

Most galaxies belong to groups or clusters, some (like our own group) containing only a few major members, some containing a hundred or more large galaxies. Examples of such rich clusters are shown in Figures 8.9 to 8.12. These have been arranged in order of increasing distance, the distances being about 400 million light years for Figure 8.9, 600 million light years for Figure 8.10, 1,250 million light years for Figure 8.11, and 3,500 million light years for Figure 8.12, according to the latest estimates of distance used by astronomers. It will be seen how, with increasing distance, the details of structure within the galaxies become more and more difficult to observe. Ultimately, at the greatest distances to which photography can be taken—about 10,000 million light years—nothing remains except a faint, soft photographic image. Galaxies at such very great distances can be seen in Figure 8.13.

FIGURE 8.4.
The Sc galaxy M51.
(Courtesy of the Kitt Peak National Observatory.)

FIGURE 8.5.
The Sc galaxy M101.
(Courtesy of the Kitt Peak National Observatory.)

FIGURE 8.6.
The "Sombrero Hat" galaxy, M104.
(Courtesy of the Hale Observatories.)

FIGURE 8.7.
The beautiful spiral galaxy M81.
(Courtesy of the Hale Observatories.)

FIGURE 8.8.
A spiral galaxy in the constellation
of Coma Berenices, seen edge-on.
(Courtesy of the Hale Observatories.)

FIGURE 8.9.
Part of the rich cluster of galaxies
in Coma Berenices. (Courtesy of
the Kitt Peak National Observatory.)

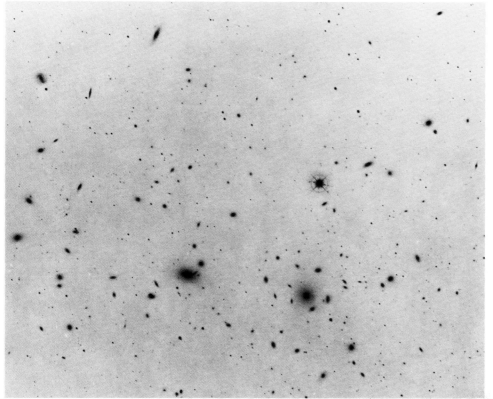

FIGURE 8.10.
Cluster of galaxies in the
constellation of Hercules,
containing many remarkable
spiral galaxies. (Courtesy of
the Hale Observatories.)

FIGURE 8.11.
Cluster of galaxies in Corona
Borealis. This cluster, intrinsically
similar to that of Figure 8.9,
is about three times more distant.
(Courtesy of the Hale Observatories.)

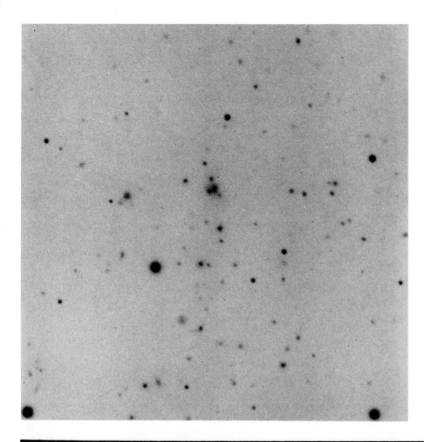

FIGURE 8.12.
In the depths of space, very many galaxies are
faintly seen. This cluster is in the constellation
of Hydra. (Courtesy of the Hale Observatories.)

FIGURE 8.13.
Galaxies between bars are at
the limit of detectability with
the largest telescopes. (Courtesy
of the Hale
Observatories.)

In Figure 8.14 we have an analysis of great importance for the study of the universe as a whole. This figure shows, in a somewhat technical way, the distribution on the sky of rich clusters of galaxies, counted to a distance of about 2,000 million light years. The figure omits the region of the sky not visible from Mt. Palomar, California, where the survey was carried out, and also omits a band along the plane of the Milky Way, because the fogging effects of dust along the Milky Way would otherwise falsify the counting of clusters in these directions. Notice that Figure 8.14 is a representation similar to that used in Figure 6.9 for displaying the Milky Way itself.

The similarity of the cluster distributions toward the two poles of Figure 8.14 implies that the universe in the large is very similar in these two opposite directions. And if the universe is similar in two such opposite directions, it is likely to be similar in *all directions*. Would the universe also appear the same if we were making our observations from some other position in space? Without actually making observations from very different positions, which obviously we are unable to do, we cannot answer this question directly. In a situation like that in Figure 8.15, the universe would appear to be the same in all directions when viewed from the center, but it would not appear the same from positions other than the center. Yet to suppose that our galaxy happens to be placed at the center of an arrangement like that in Figure 8.15 seems implausible and unsatisfactory. Rather does it seem much preferable to suppose that indeed one position in space is essentially the same as any other position in space, that the universe in the large appears essentially the same in whatever direction we look at it and from whatever parent galaxy in space we make our observations. In other words, if we had happened to live in some other quite different galaxy, situated even a large distance

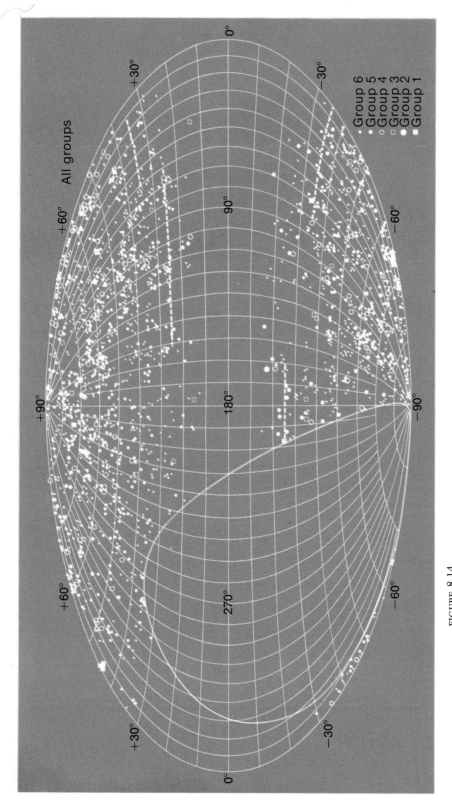

FIGURE 8.14.
The distribution of rich clusters of galaxies, with the larger symbols representing nearer clusters. (From G. O. Abell, *Astrophysical Journal*, Supplement Series, vol. 3, 1957–58, p. 211. The University of Chicago Press.)

away from our own galaxy, the universe in the large would have looked the same. This idea of "uniformity" plays an important part in the development of ideas concerning the nature of the universe as a whole.

Examination of photographs of galaxies shows that not all of them are alike. Some galaxies show patterns and others do not. Indeed, there is a whole class of galaxy, smooth in appearance with elliptical outline, and hence known as elliptical galaxies, that appear to be quite different from the spiral galaxies. The cluster shown in Figure 8.9 shows many of these elliptical galaxies, whereas the cluster of Figure 8.10 shows many spiral galaxies.

Edwin Hubble (1889–1953) classified the galaxies into various categories, illustrated by typical cases in Figures 8.16, 8.17, and 8.18. The sequence E0, E1, E2, . . . , E7, of ellipticals follows a progressive flattening of the profile, with E0 being of globular form and with E7 of a markedly flattened lenticular form. The Sa, Sb, Sc spiral forms are classified according to the importance of the central bulge, the "nucleus" as it is usually called. Spirals of type Sa have a strongly formed nucleus, those of type Sb have a somewhat smaller nucleus and those of type Sc have only a rather small nucleus. Our galaxy and the Andromeda Nebula are of type Sb, but M33 is of type Sc. Because spirals sometimes possess a central bar as well as a set, or sets of arms, Hubble also introduced the barred sequence SBa, SBb, SBc, in which the importance of the nucleus plays a similar role to that of the Sa, Sb, Sc sequence.

Galaxies falling outside this scheme were classified as "irregular." Examples of irregular galaxies are shown in Figures 8.19 to 8.24. Hubble regarded such irregular cases as being rare and not particularly important. In more recent times astron-

FIGURE 8.15.
Although different concentric shells contain different densities of galaxies, the situation is still the same in all directions for an observer at the center, but would not be so for observers not at the center.

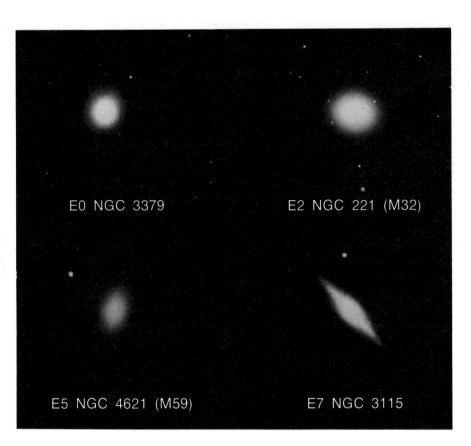

E0 NGC 3379

E2 NGC 221 (M32)

E5 NGC 4621 (M59)

E7 NGC 3115

FIGURE 8.16.
Examples of Hubble's elliptical sequence E0 to E7. (Courtesy of the Hale Observatories.)

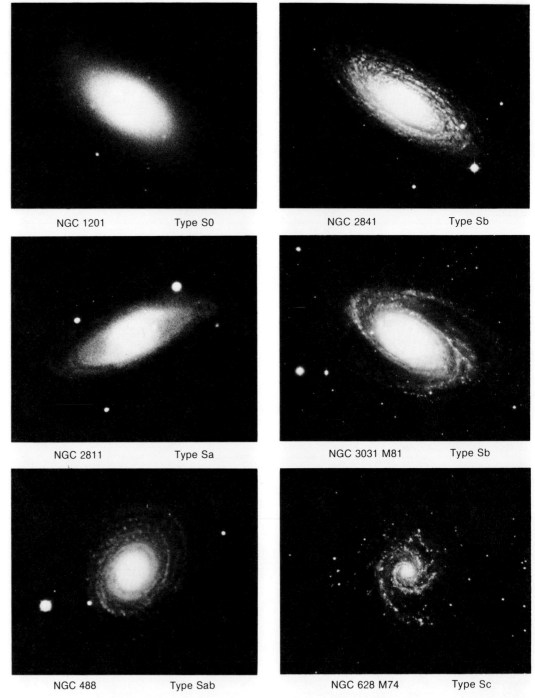

NGC 1201  Type S0

NGC 2841  Type Sb

NGC 2811  Type Sa

NGC 3031 M81  Type Sb

NGC 488  Type Sab

NGC 628 M74  Type Sc

FIGURE 8.17.
Examples of Hubble's Sa, Sb, Sc types, with an intermediate Sab form.
(Courtesy of the Hale Observatories.)

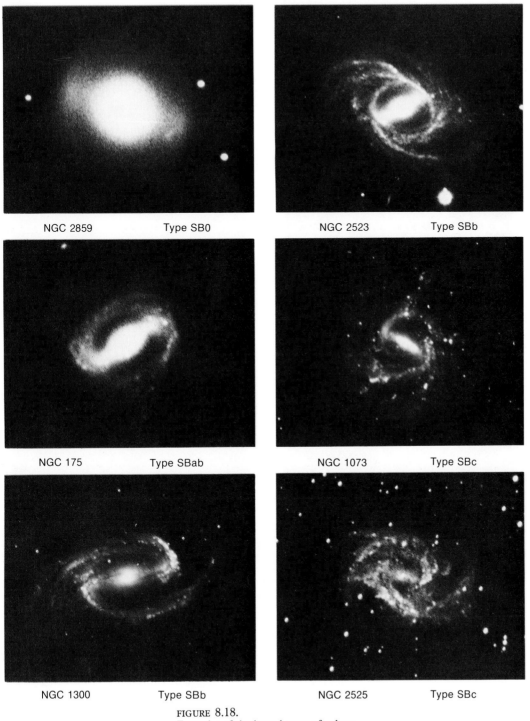

NGC 2859      Type SB0

NGC 2523      Type SBb

NGC 175      Type SBab

NGC 1073      Type SBc

NGC 1300      Type SBb

NGC 2525      Type SBc

FIGURE 8.18.
Examples of the barred types of galaxy.
(Courtesy of the Hale Observatories.)

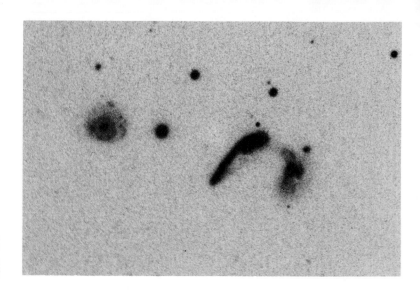

FIGURE 8.19.
A photographic negative of a peculiar chain
of galaxies. (Courtesy of Dr. H. C. Arp,
Hale Observatories.)

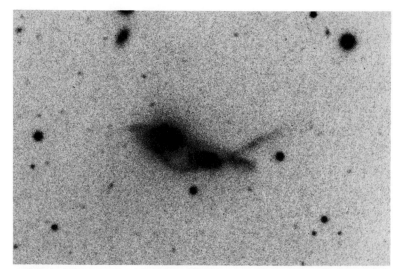

FIGURE 8.20.
Another peculiar system, south of NGC 2841.
(Courtesy of Dr. H. C. Arp, Hale Observatories.)

FIGURE 8.21.
The remarkable interacting system of galaxies
known as Stephen's quintet. It is clear that this
situation cannot persist for very long, not for a
time scale comparable to the age of our own
galaxy. It is interesting to wonder what
happened to cause this situation.
(Courtesy of the Hale Observatories.)

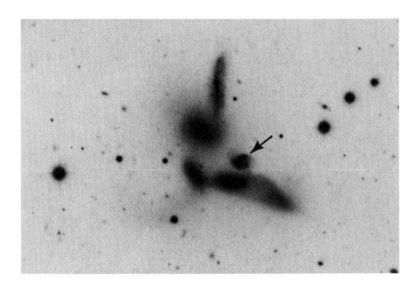

FIGURE 8.22.
Another remarkable group of galaxies, known as Seyfert's sextet. Many astronomers think the galaxy indicated by the arrow may be a background object and not a true member of the group. (Courtesy of Dr. H. C. Arp, Hale Observatories.)

FIGURE 8.23.
The strange case of a galaxy (NGC 7603) which appears to be connected to a smaller companion galaxy. Yet measurements have suggested that the companion galaxy is at a distance of 1,000 million light years, whereas the main galaxy is at a distance of 500 million light years. (Courtesy of Dr. H. C. Arp, Hale Observatories.)

FIGURE 8.24.
Another peculiar system, NGC 3561. (Courtesy of Dr. H. C. Arp, Hale Observatories.)

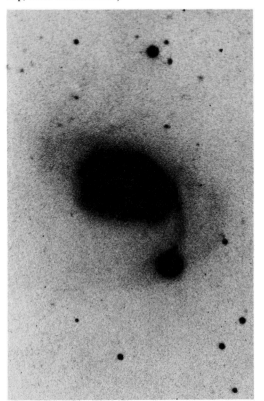

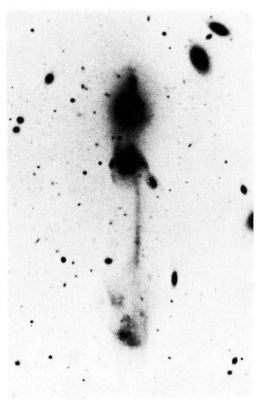

omers have become much more actively interested in these exceptional galaxies, and think that they are not quite as rare as Hubble supposed. Figure 8.25 shows the galaxy M87, which appears in this figure as an apparently normal globular galaxy belonging to the class of "ellipticals." However, on a shorter time exposure, M87 has the appearance shown in Figure 8.26. The jet which can be seen emerging from the center is by no means "normal." The light from this jet does not come from stars at all, but from very high-speed electrons moving in a magnetic field. The electrons also emit other forms of radiation, including radiowaves, so that M87 is known as a "radiogalaxy."

The jet of M87 may not be the one-sided affair it looks like in Figure 8.26; there could well be an oppositely directed jet with a somewhat different magnetic structure which does not happen to emit visible light. Such oppositely directed jets of high-speed particles are typical of huge outbursts that occur in the central regions of galaxies. When these jets impinge on an external magnetic field, or on an external cloud of gas, they cause an intense emission of radio waves. Some examples of radiogalaxies with this double pattern of emission are shown in Figure 8.27. In Chapter 6 we discussed the explosion of stars, and we saw how such explosions, known as supernovae, can cause a star to become temporarily as bright as a whole galaxy. But the outbursts from radiogalaxies are much more energetic than those of supernovae, more on the order of a million supernovae all exploding together. Further examples of radiogalaxies are shown in Figures 8.28 and 8.29.

FIGURE 8.25.
The slightly elliptical galaxy M87.
(Courtesy of the Hale Observatories.)

FIGURE 8.26.
The jet of the galaxy M87. This is
a blowup of the extreme central
region of the galaxy shown in its
entirety in Figure 8.25.
(Courtesy of the Hale Observatories.)

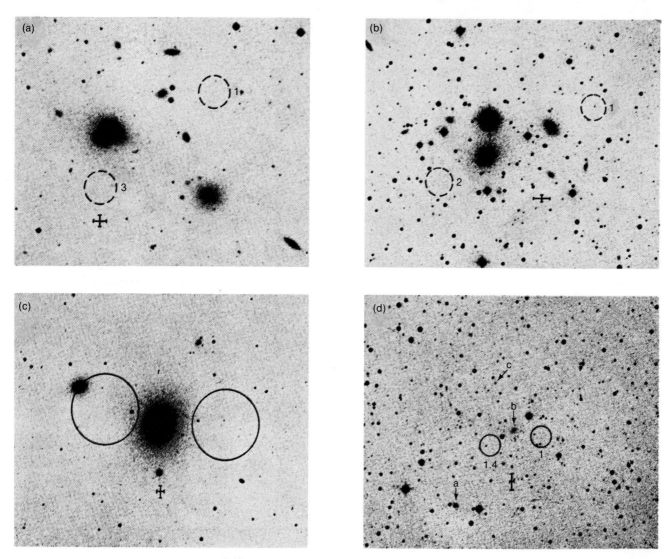

FIGURE 8.27.
Examples of radiogalaxies with double patches of radio emission.
(From P. Maltby, T. A. Mathews, and A. T. Moffet, *Astrophysical Journal,* vol. 137, 1963, p. 156. The University of Chicago Press.)

FIGURE 8.28.
This is a giant elliptical galaxy, out of which a
vast cloud of gas and dust seems to have emerged.
(Courtesy of the Hale Observatories.)

FIGURE 8.29.
One of the most powerful of all radiogalaxies,
the system known as Cygnus A.
(Courtesy of the Hale Observatories.)

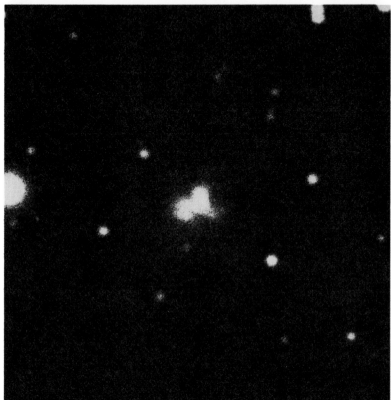

Imagine the nucleus of a radiogalaxy taken by itself. The result, for the emission of radiowaves, would be like a class of objects known as *quasars*. However, in terms of ordinary light, the quasars are thought to be often much brighter than galaxies. When photographed directly in the ordinary way, quasars appear like faint, rather blue stars, as can be seen from the examples shown in Figure 8.30. However, instead of being ordinary stars, the objects of Figure 8.30 are all thought to be situated at distances of more than 2,000 million light years away from us. Notice how the quasar denoted by catalogue number 3C273 has a jet rather like that of the galaxy M87. This jet is more clearly seen in Figure 8.31.

Why should quasars be brighter than galaxies? Start with an ordinary galaxy, and suppose that a vast cloud of high-speed electrons is generated in the central nucleus. Let these electrons, moving in magnetic fields, emit light very intensely, much more intensely than the jet of the galaxy M87. Indeed, let the nucleus become so inordinately bright that, when seen from a great distance, the nucleus overwhelms the ordinary starlight of the galaxy. Seen from a great distance, only a central brilliant point of light can then be distinguished. Many astronomers believe this to be a quasar. In short, a quasar is a galaxy in which the central nucleus has temporarily become exceedingly bright.

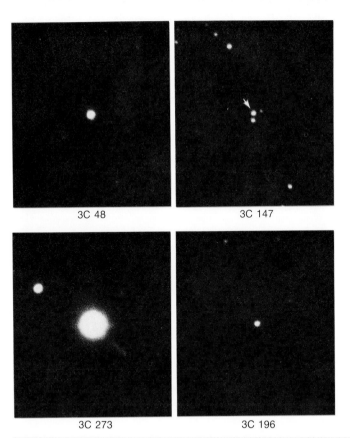

3C 48

3C 147

3C 273

3C 196

FIGURE 8.30.
When photographed directly, the quasars appear like stars.
(Courtesy of the Hale Observatories.)

FIGURE 8.31.
A long exposure in negative form shows
the jet of the quasar 3C 273.
(Courtesy of the Hale Observatories.)

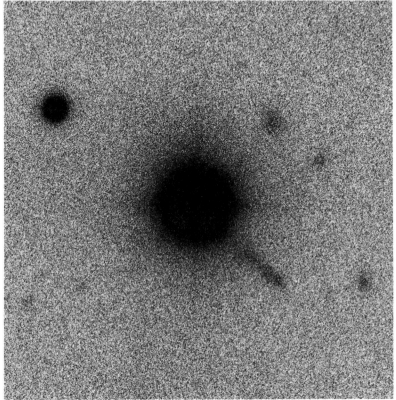

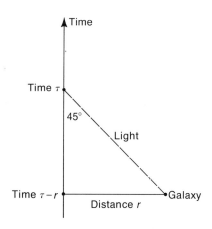

FIGURE 8.32.
Light travels from the galaxy $G$, and is received by the observer $O$ at time $t$.

We come now to the final issue to be treated in this book. Although we will be looking at matters in a rather different way than is usual, *the following discussion is in fact precisely equivalent to the usual description of the expansion of the universe.* At the beginning of Chapter 3, we discussed the meaning of the idea of "mass." To calculate the mass of a body, we counted the numbers of the various kinds of atoms which it contained. Then to each kind of atom we gave a number: 1 to the simplest form of atom, hydrogen; 12 to an atom of carbon; 16 to oxygen; 56 to iron; and so on. By adding all the resulting numbers, we found the mass of the body.

Our standard unit in such a calculation was the hydrogen atom. The question we now have to ask is whether the hydrogen atom is always the same wherever in the universe it happens to be located.

How might we know if there was a difference? A possible test for such a variation is illustrated in Figure 8.32. Light from a distant galaxy, denoted by $G$, travels across space, arriving in our telescopes at the present moment of time, which we denote by $t$. Even though light travels very fast, it still may need a long time to make its journey, just because the galaxy $G$ is very distant. In fact, if $G$ is distant by 1,000 million light years, the light will require 1,000 million years to reach us. This is such a long time ago that we can wonder whether hydrogen atoms present in the galaxy $G$ really were the same 1,000 million years ago as they are today on the Earth. How shall we know? By examining the light which the hydrogen atom then emitted. If the distant atoms had a different mass from local atoms, the light which they emitted, and which we are now receiving in our

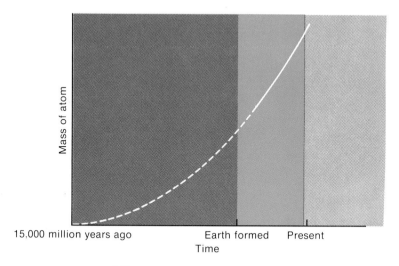

FIGURE 8.33.
The behavior of the mass of the hydrogen atom with respect to time.
The solid part of the curve is observed, the dotted part inferred.

telescopes, will have a different color from the light emitted by ordinary hydrogen atoms in a terrestrial laboratory. Here then is a direct test of the idea. Suppose we examine the galaxies of Figures 8.9, 8.10, 8.11, and 8.12 from this point of view. What is the result of such a test?

The result is that the light is different, and it is different progressively: the more distant the galaxy, the more the light is changed. The manner of change, a reddening of the light—known as the *redshift*—indicates that the mass of the hydrogen was less at earlier times than it is today. The observations indicate the behavior shown in Figure 8.33. Here the mass of the hydrogen is shown varying with time, in such a way that the earlier the time the smaller the mass. The observations so far made are represented by the solid part of the curve. The dotted part of the curve is *inferred* from our

understanding of physical laws, especially from the theory of gravitation put forward by Albert Einstein (1879-1955). This inferred part of the curve has the peculiarity that it goes back to a time when the mass was zero, nothing at all. According to our observations and calculations, this was the situation some 15,000 million years ago. This most peculiar situation is taken by many astronomers to represent the *origin of the universe*. The universe is supposed to have begun at this particular time. From where? The usual answer, surely an unsatisfactory one, is: from nothing! The elucidation of this puzzle forms the most important problem of present-day astronomy, indeed, one of the most important problems of all science.

The concept that the universe originated from nothing is illustrated in Figure 8.34, with all the particles beginning their existence at the moment

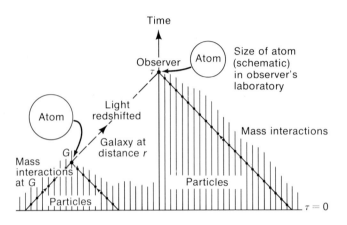

FIGURE 8.34.
This arrangement of the particle paths can be shown to be mathematically equivalent to a well-known cosmological model for the universe, with all the particles coming into existence at the moment $t = 0$.

$t = 0$. Why? It is not usual in present-day cosmological discussions to seek an answer to this question; the question and its answer are taken to lie outside the range of scientific discussion.

Yet we can hardly avoid wondering why the particles cannot be extended backward in time, as in Figure 8.35. The reason lies in the behavior of the particle masses shown in Figure 8.33, a behavior which can be understood in terms of interactions between the particles. The interactions experienced at time $t$ by a particle go along the lines in Figure 8.34 which slope at 45°, and these interactions lead to just the mass dependence shown in Figure 8.33. Since there can be no such interactions before $t = 0$, the particle masses at $t = 0$ must be zero, as they should be to agree with Figure 8.33. But the particle masses would not be

zero if we sought to use Figure 8.35 instead of Figure 8.34, for interactions would then precede $t = 0$. In fact, a calculation of the particle masses according to Figure 8.35 leads to a result that is not consistent with the part of the curve of Figure 8.33 that is based on actual observations of galaxies. Consequently the situation shown in Figure 8.35 cannot be correct.

A way to avoid this conflict with observation is shown in Figure 8.36, which is like Figure 8.35 except that interactions preceding $t = 0$ are taken to have a negative sign. The situation in Figure 8.36 is that the moment of time $t = 0$ separates positive interactions from negative ones. Without requiring the universe to begin abruptly at $t = 0$, this idea leads correctly to the behavior of Figure 8.33.

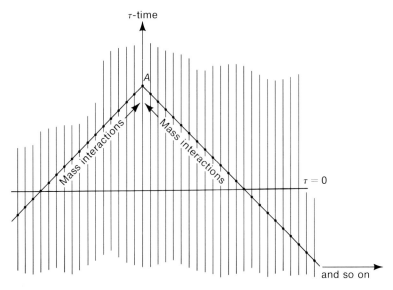

FIGURE 8.35.
Simply extending the paths of the particles backward in time leads to difficulties.

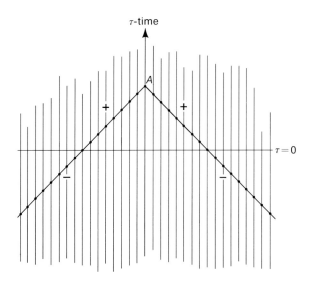

FIGURE 8.36.
The difficulties are overcome, provided the moment $t = 0$ is taken to separate positive and negative contributions to the masses of the particles.

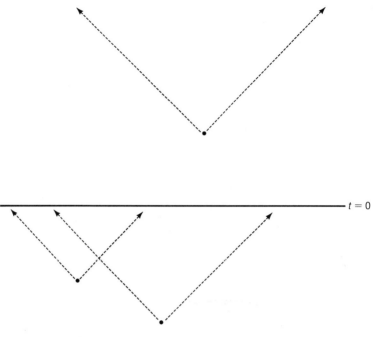

FIGURE 8.37.
Radiation propagates in the directions of the dotted lines, leading
to the origin of what is known as the "microwave background."

One important consequence of this speculation is that the universe would possess another half, a half preceding $t = 0$. Suppose that light travels in the same sense in both halves, as in Figure 8.37, and suppose that galaxies and stars exist on the "other side," just as they do on "our side." Would we then expect to be able to observe these other galaxies? Unfortunately, not directly, because all the light from the stars of such galaxies must be very strongly scattered, absorbed, and blurred at times close to $t = 0$ by particles lying between the stars. This strong blurring effect is caused by smallness of the particle masses near $t = 0$. However, the blurred radiation would continue into "our half" and would indeed be observable. Such radiation is in fact observed, and is known as the "microwave background," a background of radiation arriving at the Earth from all over the sky. The observed radiation also has the intensity to be expected from the process illustrated in Figure 8.37. The existence of this microwave radiation evidently favors this speculation quite strongly.

Another interesting aspect of these ideas is that stars and galaxies on our side turn out to have a relationship to stars and galaxies on the other side. Stars approaching $t = 0$ from the other side seem able to maintain their identity as they pass through $t = 0$. The state of affairs on our side is thus subject to a measure of control from the other side, in the sense that information concerning the other side passes to us through the moment $t = 0$. Hence we may owe much of our present world to the situation on the far side of the "surface of zero mass" which has hitherto been thought to represent the origin of the universe.

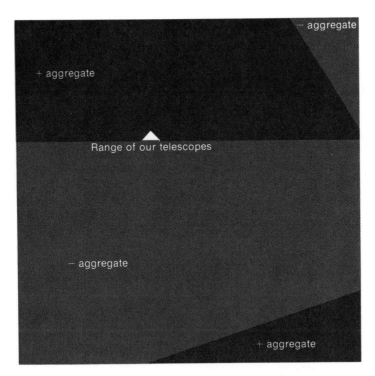

**FIGURE** 8.38.
The range of our telescopes, represented here by the white triangle, may reveal only a tiny element of the entire universe.

As a final concept, let us compare Figure 8.36 with Figure 8.38. The moment $t = 0$ of Figure 8.36 is just one of many surfaces of zero mass in Figure 8.38, and the range of our astronomical observations has become the small region indicated in Figure 8.38 by the white triangle. This generalization of our cosmological ideas suggests that the portion of the universe studied so far by astronomers may be no more than a tiny element of the whole. The full measure of all things may be vastly richer than we usually suppose.

*Questions and Topics*
*for Discussion*

## CHAPTER 1

1. Describe the appearance of the Earth as it is seen during a journey to the Moon.

2. Compare the appearance of the Earth with that of the Moon.

3. What was the Apollo program and what did it achieve?

4. Estimate the fraction of the Earth that is cloud-covered. Which regions of the Earth would you expect to be usually free from clouds?

5. Why does a color photograph of the whole Earth appear blue at its edges?

6. What do you understand by the term "plate motion?" How does plate motion affect the continents of the Earth?

7. Suppose you visited the Earth as it was 200 million years ago. In what ways would living creatures, the oceans, and the continents be different from their present-day forms?

8. Give a broad account of the evolution of life on the Earth. How old is the Earth?

9. How far do you consider mountain ranges to be the wastelands we usually take them to be? How far should violent phenomena, such as hurricanes and volcanoes, be thought of as genuine disasters?

10. Suppose that in the universe there are many planets broadly similar to the Earth, but differing somewhat, one from another, in their gravity, temperature, and atmospheric gases. How far do you consider that life, if it originates on these planets, might be similar to terrestrial life forms? Different from terrestrial forms?

## CHAPTER 2

1. The Earth's axis of rotation, instead of being perpendicular to the plane of the Earth's orbit around the Sun, is tilted at 23.5°. What effect does this tilt have on our daily lives?

2. Suppose the tilt had been 90°. What effect would this have had?

3. Explain what is meant by the calendar year, using as examples the calendars introduced by (a) Julius Caesar, (b) Pope Gregory. Why do such systems fail to keep step with the true year?

4. The nearest star is about $4 \times 10^{13}$ kilometers from us. Compare this interstellar distance with the scale of the planetary system, estimating the latter from the orbit of Neptune.

5. If you have access to a pocket calculator, make up a table similar to Table 2.1, but giving in the second column the *cube* of the average distance of each planet from the Sun (using the Earth's distance as standard) and in the third column giving the *square* of the time required for a complete orbital revolution of each planet.

6. An object attached to a piece of string is whirled around in a circle. How do you expect the pull in the string to depend on (a) the weight of the object, (b) the speed with which it is whirled, (c) the length of the string?

7. The two preceding topics can be used to investigate the nature of the Sun's gravitational pull on a planet. How? (This is a hard question, but worth thinking about—it was a major problem confronting scientists in the seventeenth century.)

8. Describe a method for constructing an ellipse. What are the foci of an ellipse?

9. Describe a major contribution to astronomy made by (a) Kepler, (b) Newton.

10. In what important philosophical respect do the views concerning the universe held by modern scientists differ from those of the ancient Greeks?

## CHAPTER 3

1. How is the mass of a body calculated in terms of the hydrogen atom as the unit?

2. Discuss the difference, both in mass and in the materials of which they are composed, between the four inner planets and the four major outer planets, Jupiter, Saturn, Uranus, and Neptune.

3. Discuss the implications of the altimeter record of the Moon (Figure 3.8).

4. Estimate the distance across Mare Orientale.

5. What are the similarities and differences between Venus and the Earth?

6. Describe the surface features of Mars, and discuss the ways in which they are subject to change.

7. Describe the appearance of Jupiter. What is known about the interior of Jupiter?

8. Describe the appearance of Saturn. How many satellites does Saturn have?

9. What chemical substances are found in the atmospheres of the four major outer planets?

10. If you were sending a message to an intelligent creature on another planet moving around some distant star, how would you attempt to convey information about the general lay-out of our own system of planets?

## CHAPTER 4

1. The radius of the Sun is about 700,000 kilometers. What is the ratio of the radius of the Earth's orbit (taken as circular) to the radius of the Sun?

2. Light travels at a speed of about 300,000 kilometers per second. How long does it take for light to travel from the Sun to the Earth?

3. The planets shine by reflecting sunlight, whereas the Sun shines by generating its own energy. How does it do so?

4. Describe the internal structure of the Sun.

5. Describe sunspots.

6. What are (a) prominences, (b) flares?

7. What is the solar wind?

8. Discuss the evidence for the existence of a magnetic field in the atmosphere of the Sun.

9. In what crucial respect does a mass of gas having internal temperature variations differ from a gas with a precisely uniform internal temperature?

## CHAPTER 5

1. Describe the process whereby the four inner planets are thought to have been formed. Why were the initial surfaces of these planets likely to have been densely cratered?

2. What similarities and what differences would you expect there to have been in the process of formation of the four major outer planets?

3. What are asteroids, and where is the asteroidal belt?

4. How does material from the asteroidal belt reach the Earth?

5. If an asteroidal lump of material some 10 kilometers in diameter were to strike the Earth, what would be the physical consequences? Estimate the probability of such an incoming missile striking a heavily populated area of the Earth.

6. Discuss the nature and properties of comets. In particular, how is the "tail" of a comet formed?

7. Describe the history of Halley's comet.

8. How do comets reach the ends of their lives?

9. What are (a) meteorites, (b) meteors?

10. As a project, find how many meteorites are discovered on the average each year. How are meteorites named?

---

CHAPTER 6

1. Compare the star maps in Figures 6.1 to 6.6 with actual observations of the sky, familiarizing yourself with as many constellations as you can identify.

2. Discuss the properties of the Orion Nebula, and describe the way in which stars are formed within gas clouds like the Orion Nebula.

3. What is a light year?

4. Why does the material within a condensing protostar become hot? Which nuclear process first becomes important within a condensing protostar?

5. Why are there color variations within the Andromeda Nebula, from orange-red in the central regions to blue in the outer regions?

6. Why are stars born in clusters? Describe the system of Castor.

7. As the hydrogen near the center of a star becomes exhausted, by being converted to helium, what happens to the star?

8. Discuss the history of the Crab Nebula.

9. Describe the nature of (a) a planetary nebula, (b) a white dwarf, (c) a supernova, (d) a neutron star, (e) a pulsar, (f) a black hole.

10. How are the materials of our everyday world believed to have originated?

---

CHAPTER 7

1. What chemical substances are thought to be important for the origin of life?

2. Given these substances, what are the main steps whereby life is thought to arise and to evolve?

3. What are the critical differences between plants and animals?

4. The energy consumption by human species is very much larger than the energy consumption by any other animal. Do you consider this to be the most objective way to describe difference between humans and other animals? If not, what other criterion would you suggest?

5. What sources of energy are available to the human species on a time scale of (a) a few decades, (b) a few centuries, (c) 10,000 years, (d) 1,000,000 years?

6. What is Project Cyclops?

7. Discuss the possibility of establishing communication with other creatures living in distant parts of our galaxy.

8. Do you think it inevitable that mankind will survive the difficult social problems that appear to be arising at the present time?

9. If your answer to the preceding question was affirmative, sketch what you think might be the course of history during (a) the next 100 years, (b) the next 1,000 years, (c) the next 10,000 years.

10. Scientists generally consider that there is no worthwhile evidence that intelligent creatures from other planets have visited the Earth, in either the recent or the remote past. This lack of evidence could be due to one of the following reasons.
    (1) Physical travel from one stellar system system to another is impossible.
    (2) We are alone in the universe.
    (3) Although physically capable of making such a visit, beings who are of higher intelligence than humans would not do so, because they would have an ethic against interfering with beings of lower intelligence, such as ourselves.
    (4) Visits have indeed taken place, but in a form we cannot recognize, just as an ant could not recognize a human.
    Discuss these possibilities (and any others that might occur to you), and explain which you consider to be the most plausible.

11. During the past 500 years, the human species has changed from living at comparatively low levels of both population and technology to living at high levels of both. How far do you consider that modern social problems would be eased by the combination of high technology and low population?

---

CHAPTER 8

1. Describe the group of galaxies of which our own galaxy is a member. What is the scale of this local group?

2. Describe the characteristics of rich clusters of galaxies.

3. Discuss evidence for the view that the large-scale structure of the universe is uniform.

4. How did Hubble classify the galaxies?

5. What are (a) radiogalaxies, (b) quasars?

6. What is the evidence for an "expansion of the universe"?

7. How may the expansion of the universe be explained in terms of a time-variability of particle masses?

8. Discuss the "origin" of the universe. How far do you consider this concept to be satisfactory?

9. How can the need for an origin of the universe be avoided?

10. Discuss the observable consequences of your answer to the preceding question, especially in terms of the origin of the microwave background radiation.

# Index

Ammonia
    in atmospheres of outer planets, 64
    as basic material for origin of life, 128
Animals
    electronic systems of, 134
    emergence of intelligence in, 135
    persistence of intelligence in, 137
    weapon systems of, 135
Apollo missions, 2
Asteroids, 83

Bernoulli, Johann (1667-1748), 34
Betelgeuse, 115
Black holes, 123

Calendar, 28, 29
Castor system, 114
Comets, 84, 88
    in collision with Earth, 85
    final breakup of, 92
    graveyard of, 91
    Halley's, 89
    tails of, 88
Copernicus, Nicolaus (1473-1543), 31
Crab Nebula, 124

Dirac, Paul (1902-    ), 36

Earth
    atmosphere and sky of, 21, 22
    continental movements of, 14
    earthquakes, 18
    evolution of life on, 7-8
    life 3,000 million years ago on, 11
    nuclear energy generated inside, 14
    orbit around Sun, 26
    origin of mountains, 17
    plate movements of, 17
    rotation of, 26
    seasons of year, 28
    seen from outside, 2, 6
    volcanoes on, 18
Eclipses of Sun, 74

Galaxies
    Andromeda Nebula, 112
    clusters of, 144
    distances of, 144
    Hubble's classification of, 152
    Large Magellanic Cloud, 140
    local group of, 140
    mass of, 108
    Messier (M) 33, 140
    Messier (M) 87, 158
    nuclei of, 162
    radio, 158
    redshifts of, 164 *et seq.*
    Small Magellanic Cloud, 140
    supernovae in, 158
Galileo Galilei (1564-1642), 67

Halley's comet, 89

Intelligence. *See* Animals
Interstellar communication, 136

Jupiter, 59
    gases in atmosphere of, 64
    excess infrared radiation from, 64
    influences cometary orbits, 88
    *See also* Planets

Kepler, Johannes (1571-1630), 33

Leibniz, Gottfried Wilhelm (1646-1716), 34
Life
    electrical effects associated with complex
        forms of, 134
    favorable sites for, 128
    on other planets, 10
    origin of, 128
    *See also under* Earth
Light, speed of, 106

Mars, 56
    dust storms on, 52
    wind erosion on, 52
    volcanoes on, 57
    *See also* Planets
Mass
    nature of, 38
    of galactic nebulae, 108
    of Moon, 40
    of planets, 39

    of Venus, 49
    time variability of, 164
Mercury. *See* Planets
Meteorites, 83
Meteors, 93
Methane
    as basic material for origin of life, 128
    in atmospheres of outer planets, 64
Microwave background radiation, 168
Mira, 116
Moon, 40
    elevation differences on, 47
    far side of, 43
    impact of debris on, 82
    Mare Orientale, 47
    phases of, 6
    unusual surface features of, 43

Nebulae
    dark, 108
    *See also* Galaxies *and by name*
Neptune. *See* Uranus and Neptune
Neutron stars, 123
Newton, Isaac (1643-1727), 34

Origin
    of chemical elements, 124
    of life, 128
    of microwave background radiation, 168
    of universe, 165
Orion Nebula, 104

Pictographs, North American Indian, 122
Planets
    composition of, 39
    distances of, from Sun, 30
    masses of, 39
    orbits of, 30, 33
    outer, 59, 64
    periods of revolution of, 30
    *See also by name*
Pulsars, 123

Quasars, 162

Rosette Nebula, 106

Saturn
    composition of, 59
    rings of, 64
    rotation of, 59
    *See also* Planets
Scientific discovery, nature of, 35
Speed of light, 106
Stars
    Betelgeuse, 115
    binary, 114
    Castor, system of, 114
    clusters of, 114
    colors of, 111
    evolution of, 115
    formation of, 110, 114
    helium-burning inside, 116
    lifetimes of, 111

    Mira, 116
    nuclear energy generated inside, 111
    on other side of universe, 168
    pulsating, 116
    white dwarfs, 117
Sun, 71
    corona of, 78
    eclipses of, 74
    flares on, 74
    gravitational influence of, 34, 66
    internal structure of, 78
    mechanical energy from interior of, 79
    nuclear energy from, 67
    particle streams from, 74
    size relative to planets, 66
    sunspots on, 71
    x-rays from, 74
Supernovae, 120, 158

Thermodynamics, 79

Universe
    homogeneity and isotropy of, 150
    microwave background of, 168
    observable region as fragment of, 169
    origin of, 165–166
Uranus and Neptune, 31, 59

Venus, 49
    surface features of, 51
    *See also* Planets
Volcanoes, 18, 57

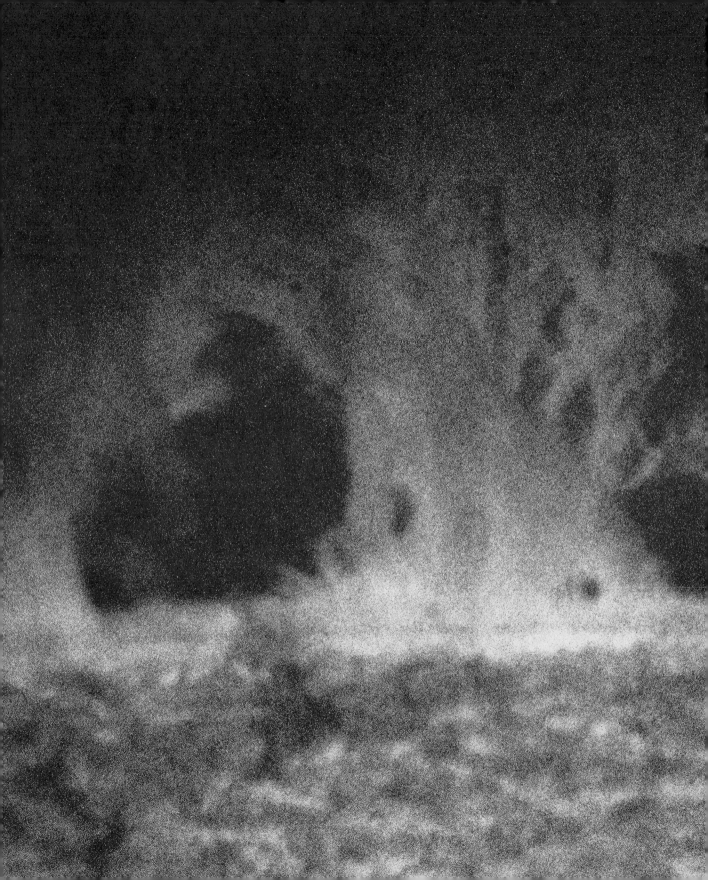